Improving Indoor Air Quality Through Design, Operation and Maintenance

Improving Indoor Air Quality Through Design, Operation and Maintenance

By Milton Meckler

Published by
THE FAIRMONT PRESS, INC.
700 Indian Trail
Lilburn, GA 30247

Library of Congress Cataloging-in-Publication Data

Meckler, Milton.
    Improving indoor air quality through design, operation, and maintenance /
by Milton Meckler.
    p. cm.
    Includes index.
    ISBN 0-88173-208-7
    1. Indoor air pollution. 2. Air quality management. I. Title.
TD883.17.M43   1996   697.9--dc20   95-49207
                                                 CIP

Improving Indoor Air Quality Through Design, Operation, And Maintenance
By Milton Meckler.

Published by The Fairmont Press, Inc.
700 Indian Trail
Lilburn, GA 30247

Printed in the United States of America

10 9 8 7 6 5 4 3 2 1

ISBN 0-88173-208-7   FP

ISBN 0-13-231820-2   PH

While every effort is made to provide dependable information, the publisher, authors, and
editors cannot be held responsible for any errors or omissions.

Distributed by Prentice Hall PTR
Prentice-Hall, Inc.
A Simon & Schuster Company
Upper Saddle River, NJ 07458

Prentice-Hall International (UK) Limited, London
Prentice-Hall of Australia Pty. Limited, Sydney
Prentice-Hall Canada Inc., Toronto
Prentice-Hall Hispanoamericana, S.A., Mexico
Prentice-Hall of India Private Limited, New Delhi
Prentice-Hall of Japan, Inc., Tokyo
Simon & Schuster Asia Pte. Ltd., Singapore
Editora Prentice-Hall do Brasil, Ltda., Rio de Janeiro

Table of Contents

Foreword

Ever since the "Clean Air Act" was passed by the U.S. Congress in 1970, billions of dollars have been spent to control dangerous emissions outdoors. Yet, indoor air contaminants (especially in our office buildings) may pose the greatest threat to human health. Concentrations of harmful contaminants are much higher indoors than outdoors. As a rule of thumb, higher indoor air contaminant concentrations can mean higher health risks. According to a growing body of scientific evidence, serious acute and chronic health risks may very well result from inadequate indoor air quality (IAQ).

What is IAQ and why is it important to us? IAQ has been described as the result of a complex relationship between the contamination sources in a building, the ventilation rate, and the dilution of indoor air contaminant concentrations with outside air. This relationship is further complicated by outside air sources entering our buildings and used for dilution air and pollution sinks that may modify or remove contaminants. This definition of IAQ must also include occupant comfort, energy use, lighting, acoustics and related "stressor" issues.

Emissions from indoor contamination sources in a building are often the primary determinant of IAQ. They include building materials, consumer products, cleaners, furnishings, combustion appliances and processes, biological growth, building occupants, etc. Since we spend most of our time indoors, contamination sources can have profound effect on our health and comfort. Inadequate IAQ today can result in absenteeism and reduced productivity in the workplace, and higher medical costs. In the light of these facts, building owners and managers are understandably concerned about today's rapidly increasing IAQ litigation. Accordingly, heating, ventilating and air-conditioning (HVAC) practitioners, building owners, operators and managers must recognize their joint responsibilities in providing for and maintaining acceptable IAQ.

"Dilution is the solution" often leads to energy waste and can be counterproductive. To minimize health risks while still providing for energy efficiency requires use of appropriate filtration/air-cleaning systems. These systems can also help reduce outside ventilation airflow rates, often resulting in significant energy savings while still providing acceptable IAQ.

In addition, the essential role of operation and periodic maintenance (preventive measures) must not be overlooked. Properly implemented operation and maintenance procedures can avoid inadequate IAQ, ensure proper operation and longevity of the HVAC system, while eliminating excessive energy consumption as will be seen.

Considering the interactive aspects of design, operation and maintenance in providing acceptable IAQ, we brought together authors who are recognized experts in their respective fields in a collaborative effort. IMPROVING INDOOR AIR QUALITY THROUGH DESIGN, OPERATION AND MAINTENANCE is a consolidated book intended for use by architects; HVAC practitioners; contractors; and building owners, operators and managers, to provide a healthy and productive indoor environment. In short, this comprehensive reference book will provide useful guidance on how to achieve acceptable IAQ by design, operation and maintenance without sacrificing energy efficiency.

While IMPROVING INDOOR AIR QUALITY THROUGH DESIGN, OPERATION AND MAINTENANCE places special emphasis on practical measures that can be easily implemented in new as well as old buildings, it also contains some state-of-the-art techniques and challenging new methods, some of which were only recently tested in the laboratory. Although the book addresses a broad range of topics in IAQ, for practical purposes we divided it into four sections.

Section 1: Introduction to Indoor Air Quality (Chapters 1 and 2) reviews the types and potential sources of indoor air contaminants, their health risks based on most recently developed data, dynamic evaluations of indoor air contaminant concentration at part-load, while complying with the requirements of ANSI/ASHRAE Standard 62-1989 in achieving acceptable IAQ.

Section 2: Monitoring and Measuring Indoor Air Contaminant Concentrations (Chapters 3 through 5) deals with the development of direct and indirect methods to verify compliance. Such methods include carbon dioxide and carbon monoxide methods. Also included in this section is the monitoring of HVAC system performance through measurements employing tracer-gases. Such quantitative measurements include air-exchange rate and age-of-air in occupied spaces.

Section 3: Practical Design Solutions that Improve Indoor Air Quality (Chapters 6 through 11) covers state-of-the-art engineering designs that can be easily employed in new buildings as well as retrofitted into the existing buildings. In addition, this section provides retrofitted system designs only recently tested for performance. It also provides guidance on how one can easily integrate such a system without any substantial cost. This section also explores the very important role of filtration (particulate and gas-phase) and air-cleaning techniques capable of controlling volatile organic compounds (VOCs), systems that employ liquid and solid desiccants to independently control temperature and humidity while achieving improved IAQ and enhanced occupant comfort, and co-adsorption of indoor air contaminants by single-component monolayer and multilayer models. Also included are liquid-desiccant enhanced heat pipe air-conditioners, and liquid/solid desiccant-assisted package and outside air preconditioning desiccant modules. Each chapter includes several examples for better demonstration purposes.

Finally, Section 4: Operation and Maintenance Procedures to Improve Indoor Air Quality (Chapters 12 through 14) deals with perhaps the most often overlooked preventive aspects of IAQ such as proper system operation and maintenance. This section offers useful strategies to maintain acceptable IAQ during building renovations; proactive IAQ building management through preventive measures such as training, compliance with applicable latest standards, adequate filtration, biocontaminant control; and proper handling of possible IAQ-related complaints to avoid costly IAQ litigation.

Sincere thanks are due to our contributors who shared their valuable time, immense knowledge and patience to make this a success that we hope will be. I would also like to express my sincere thanks to Refik A. Sar, Director of Communications at The Meckler Group for his continuous and invaluable assistance to me in the preparation of this material.

Milton Meckler, P.E.
President
The Meckler Group

Contributors

H.E. Barney Burroughs, IAQ/Building Wellness Consultancy, Chapter 6.

S.W. Choi, Ph.D., Illinois Institute of Technology, Chapter 8.

Francis M. Gallo, P.E., CFM, CIAQP, LZA Technology, Chapters 12 and 13.

Tushar K. Ghosh, Ph.D., University of Missouri - Columbia, Chapter 10.

Anthony L. Hines, Ph.D., P.E., Honda of America Mfg., Inc., Chapter 10.

Larry C. Holcomb, Ph.D., Vice-President, Holcomb Environmental Services, Chapter 1.

Milton Meckler, P.E., President, The Meckler Group, Chapters 1, 2, 3, 4, 5, 7, 9 and 11.

Philip R. Morey, Ph.D., CIH, AQS Services, Inc., Chapter 14.

Demetrios J. Moschandreas, Ph.D., Illinois Institute of Technology, Chapter 8.

Davor Novosel, Gas Research Institute, Chapter 9.

Yves Parent, Ph.D., National Renewable Energy Laboratory, Chapter 9.

Ahmad Pesaran, Ph.D., National Renewable Energy Laboratory, Chapter 9.

Section 1

Introduction to Indoor Air Quality

1

IAQ — Definitions and Sources of Indoor Air Contaminants

Milton Meckler, P.E.
President, The Meckler Group
Encino, California

and

Larry C. Holcomb, Ph.D.
Vice-President, Holcomb Environmental Services
Olivet, Michigan

1.1   Introduction
1.2   Indoor Air Contaminants and Their Sources
1.3   References

1.1   INTRODUCTION

Since most of our time is spent indoors, indoor air quality (IAQ) is very important to our health and comfort. IAQ is a result of a complex relationship between the contamination sources in a building and the ventilation rate. This relationship is further complicated by outside sources used for dilution air and pollution sinks in a building that may modify or remove contaminants. Some of the major factors influencing IAQ in a building may be summarized as: (a) emissions from indoor contamination sources, (b) quality of the outside dilution air, (c) dilution rate of outside ventilation air, and (d) systems and materials in a building that change the concentrations of contaminants.

Emissions from contamination sources in a building are the primary determinant of IAQ. They include building materials, consumer products, cleaners, furnishings, combustion appliances and processes, biological growth, building occupants, etc. The physiological reactions to the air contaminants, coupled with the psycho-social stresses of the modern office environment and the wide range of human susceptibility to indoor air contaminants, have led to acute building sicknesses such as sick building syndrome (SBS), building-related illness (BRI), and multiple chemical sensitivity (MCS).

SBS is often associated with buildings in which a majority of occupants experience a variety of health and comfort problems for which no specific cause can be identified. Health-related complaints from occupants often include irritation of the eyes, nose, throat and upper respiratory system; headaches; and general fatigue. Because of limited data to date, quantitative information has also been limited. However, while data are insufficient to provide quantitative determination of health risks, the available data serve as an indicator of the potential risks associated with IAQ. On the other hand, BRI refers to an illness caused by exposure to an indoor air agent that can be identified. Symptoms include infection, fever, hypersensitivity, pneumonitis, etc.

Anecdotal data suggest the possibility that a small group of the population has become sensitive to chemicals in the environment. These individuals appear to suffer acute reactions, collectively known as MCS, repeatedly upon exposure to levels commonly encountered in indoor environments that would not affect most people. Although such individuals report significant adverse effects, the lack of clear diagnostic criteria, data or tools to evaluate the extent of the effects often leads to some skepticism. It is not known to what extent MCS affects occupant health or productivity or what the size of this population affected by MCS is. It is a subject of considerable intra-professional disagreement concerning the existence and etiology of this potential illness. While no widely accepted test of physiologic function has been shown to correlate with the reported symptoms, the sheer mass of anecdotal data is reasonable cause for concern. In this introductory chapter to IAQ, we will summarize some of the most important indoor air contaminants and their sources. These include radon, formaldehyde, particulates, products of combustion, volatile organic compounds (VOCs) and environmental tobacco smoke (ETS).

1.2   INDOOR AIR CONTAMINANTS AND THEIR SOURCES

One of the most important indoor air contaminants is formaldehyde. A member of a chemical group called aldehydes, formaldehyde is highly reactive and soluble. It may exist as a pure gas, an aqueous solution, and a solid polymer[1]. It is highly irritable to moist body surfaces. Urea-formaldehyde-foam insulation (UFFI), particleboards, some paper products, fertilizers, chemicals, glass and packaging materials are the major sources of formaldehyde. Formaldehyde has been identified as a "carcinogen" and linked particularly to nasal cancer.

A recent review of indoor air contaminant concentrations[2], including formaldehyde, has been published. Although some office environments may be exposed to higher concentrations of formaldehyde due to off-gassing, concentrations found were typically lower than those of homes (0.008 parts per million [ppm] to 0.04 ppm). Concentrations of formaldehyde measured in homes ranged from 0.02 ppm to approximately 4.2 ppm in a mobile home with IAQ complains in Wisconsin. Mobile homes have been found to have higher concentrations of formaldehyde because of the extensive use of particleboards and plywood in construction, tight sealing, and the increased use of recirculated air.

Another significant source of indoor air contamination is the radon gas. Individuals have been exposed to natural radiation because of the radioactive elements contained in the bedrock and soils. All rocks and soils contain these radioactive elements such as uranium and thorium in very low concentrations. Exposure to higher concentrations of these radioactive elements may have serious health effects. Radon, a decay product of radium-226, is an odorless, colorless, radioactive gas and it is chemically inert. Radon decays into four short-lived daughter products. When they decay, two radon daughter products with a short half-life of approximately 30 minutes emit alpha particles and the other two daughter products emit beta particles[3].

The main source of indoor radon is the soil. It enters the building through open sumps, crawl spaces, hollow concrete block walls, and cracked concrete slabs. It also enters the building from the outside air and building materials such as concrete, brick, and stone or drywall made of phosphogypsum[3]. In multistory buildings, the main source of radon is the building materials[4]. The main health hazard due to exposure to radon is the increased possibility of cancer of internal organs such as lung cancer.

Small particulates penetrating into the lungs may cause various health problems. Asbestos, used for insulation and fireproofing, has become a significant IAQ problem. Major sources of asbestos include soils, asbestos fibers brought in by workers of asbestos mines and factories, and surface and finishing materials. Asbestos contamination occurs when the fibers become friable and released into indoor air[5]. Asbestos is considered to be a carcinogen and has been linked to stomach and lung cancers, cancer of the pleura or peritoneum, and fibrotic lung diseases[4]. Fiberglass and rockwool are the other insulation materials that may cause health problems indoors. Although the carcinogenicity of fiberglass and rockwool is currently being investigated, they are considered to be much less carcinogenic than asbestos.

Throughout our lives, we are exposed to many microorganisms. These range from viruses, bacteria, and fungal spores to pollen grains and dust mites. Some of them cause sneezing and wheezing and others may cause severe diseases. They can grow and multiply rapidly under the right conditions causing discomfort to especially sensitive people. Many materials used in building construction and furnishings and in food products are nutrient sources to microorganisms. Additionally, a contaminated heating, ventilating and air-conditioning (HVAC) system can distribute millions of microorganisms into our offices or residential buildings. Every day we consume many things produced by microorganisms. For example, enzymes incorporated into detergents are produced by fungi or bacteria. Citric acid added to many food products and antibiotics used to treat many diseases are also produced by microorganisms.

Microorganisms indoors are of great concern to us, because they can cause many diseases. Some microorganisms can cause allergic reactions, irritation and weaken immunity so that our bodies may become more susceptible to other diseases. High humidity and condensation also increase the number of airborne fungal spores and bacterial growth. While most bacteria are 0.5 μm to 2.0 μm in size, the size of fungal spores can range from 2.0 μm to 150 μm. When inhaled, fungal spores cause asthma, rhinitis and other allergic symptoms in humans. The mycotoxins produced by fungi may be poisonous, carcinogenic or immunosuppressive. However, not all fungi produce mycotoxins. Although the importance of mycotoxins in indoor environments need further evaluation, some problems related to them have been reported.

Bacteria play an important role in our environment and health. Bacteria have been used to produce chemicals such as enzymes and organic acids. Protein particles and enzymes produced by bacteria are allergenic. Bacteria found in indoor environments can be basically divided into two groups: (a) environmental and (b) human-related. Human-related bacteria include *Streptococcus*, *Staphylococcus* and some *Micrococcus*. Some of the environmental bacteria include *Basillus*, *Methylobacterium* and

Pseudomonas. Most bacteria grow in water or in substrate with high water content. They also grow on human skin and may be shed into air. While most bacteria are not pathogenic, some bacteria are opportunistic pathogens.

Byproducts of combustion are another source of indoor air contamination. Among the major byproducts of combustion are carbon monoxide (CO), sulfur dioxide (SO_2), nitrogen dioxide (NO_2), carbon dioxide (CO_2) and ETS. Major sources of combustion byproducts indoors are wood stoves, fireplaces in homes, inverted space heaters, gas stoves, tobacco smoke, etc. Vehicle exhaust and fossil-fuel-burning plants are among the major outside sources of combustion byproducts.

CO is an odorless, tasteless and colorless gas. It is released during the combustion of carbon-containing materials in an oxygen-deficient environment. It combines with the hemoglobin in the blood to form carboxyhemoglobin (COHb) which interferes with the ability of blood to carry oxygen to the body tissues[6]. High-level exposure to CO may cause dizziness, headaches, nausea, vomiting, coma, and even death. Table 1-1 shows the summary of CO concentrations in ppm measured in a variety of studies in smoking and nonsmoking areas under real-life conditions[2]. The data show mean levels of CO concentrations in nonsmoking areas ranging from 1.3 ppm (trains) to 11.6 ppm (automobiles). The mean levels in smoking areas range from 2.2 ppm (trains) to 6.4 ppm (taverns/bars). The mean difference in CO concentrations between smoking and nonsmoking areas where data are available appears to be negligible (1.0 ppm to 2.0 ppm).

SO_2, a colorless gas, is produced during the combustion of sulfur-containing materials. It is highly soluble in water and absorbed easily by the mucous membranes of the respiratory system. When inhaled, it dissolves and forms sulfuric acid, sulfurous acid and bisulfate ions[6]. Exposure to SO_2 primarily affects the respiratory system. A concentration of approximately 5 ppm starts affecting healthy human beings[7].

Exposure to NO_2, a dark brown gas, results in severe irritation to eyes and other membranes of the body, and damage to the lungs. Typical indoor concentrations of NO_2 have been reported[2]. Thirteen studies were reviewed, ten in the home environment and three in the office environment. Most of these studies have attempted to correlate fluctuations in NO_2 concentrations with the presence or absence of various appliances such as gas stoves or kerosene heaters. Concentrations of NO_2 reported in homes tend to range from 10 µg/m³ to 50 µg/m³ (19 parts per billion [ppb]

Table 1-1. Concentrations of Carbon Monoxide in Smoking and Non-smoking Areas for Several Occupancies.

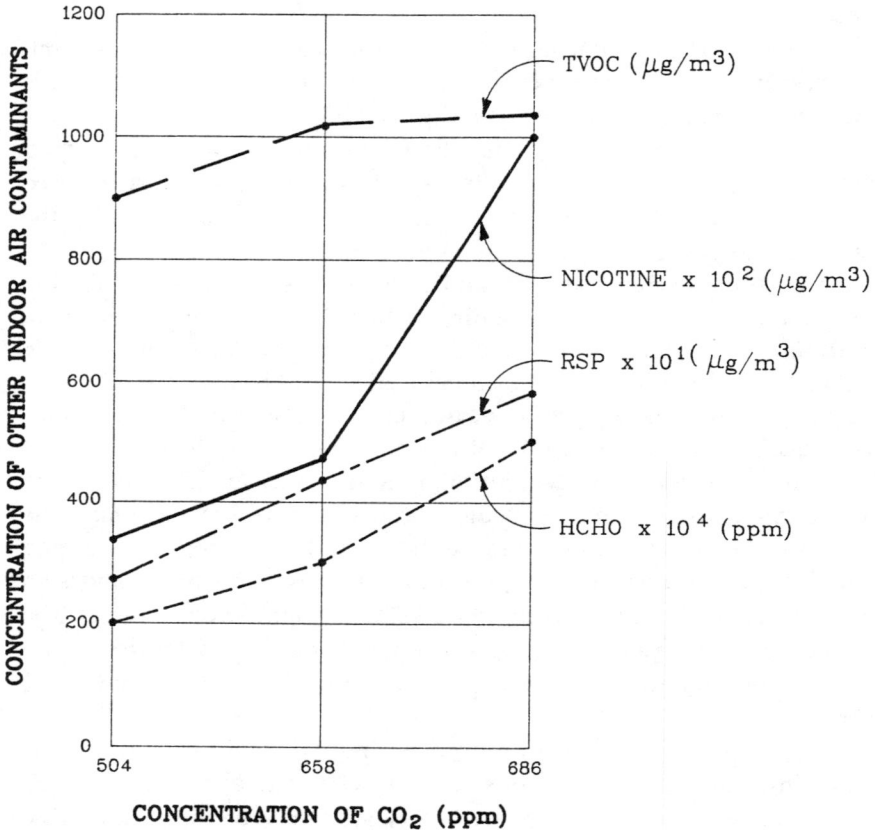

CONCENTRATION OF OTHER INDOOR AIR CONTAMINANTS

TVOC (μg/m^3)

NICOTINE x 10^2 (μg/m^3)

RSP x 10^1 (μg/m^3)

HCHO x 10^4 (ppm)

CONCENTRATION OF CO_2 (ppm)

LEGEND
```
TVOC : TOTAL VOLATILE ORGANIC COMPOUNDS
RSP  : RESPIRABLE SUSPENDED PARTICULATES
HCHO : ASBESTOS
```

to 95 ppb) with a small number of homes having high concentrations because of improperly maintained or operated gas/kerosene appliances.

CO_2, another combustion byproduct, is relatively nontoxic and considered to be an excellent surrogate for odor and indirect measure of the adequacy of mixing outside and recirculated air. Based on the rationale for minimum physiological requirements for respiration air developed in

Appendix D of ASHRAE Standard 62-1989, it is recommended that the concentrations of CO_2 not exceed 1000 ppm, which corresponds to a ventilation airflow requirement of 15 cfm per person. In Chapter 3, we will develop an indirect method utilizing measured CO_2 concentrations as a sensing control strategy to maintain adequate ventilation rates.

Recently, literature reporting measurements of CO_2 and other IAQ parameters, including ventilation rates, was reviewed. Concentrations of a variety of indoor air contaminants were compared with those of CO_2 in a number of different settings. Trends in the data indicate that concentrations of respirable suspended particles (RSP), VOCs, formaldehyde, and nicotine generally tend to correlate with varying CO_2 concentrations. Thus, using CO_2 as a surrogate may be beneficial in maintaining adequate IAQ if one does not place too much dependency on this one method alone. Figure 1-1 shows the relationship between the concentrations of CO_2 and the above-mentioned indoor air contaminants[8].

VOCs are used in the formulation or manufacturing of almost all materials and products such as construction materials, furnishings, fuels, consumer products and pesticides. Some of the commonly used pesticides include chlordane, heptachlor, malathion, diazinon, dursban, ronnel and dichlorvos. Indoor VOC concentrations are often much higher than outside concentrations. They are lipid soluble compounds and easily absorbed through the lungs. Exposures to VOCs are often multiple and symptoms are common to many compounds. There are two major types of VOCs: (a) petroleum-based solvents and (b) chlorinated solvents. The petroleum-based solvents are generally found in products such as paints, stains, adhesives and caulks. Chlorinated solvents are common ingredients of shoe polishes, water repellents, epoxy paint sprays, paint removers and dry-cleaning compounds. Some of these chlorinated solvents are carbon tetrachloride, dichloroethane, trichloroethane, dichloroethylene, trichloroethylene, tetrachloroethylene, etc.[9]

Several studies in which the VOC and total volatile organic compound (TVOC) concentrations measured are performed and reported in recent publications[10, 11]. Table 1-2 shows the locations, number of studies conducted, number of locations, and the average TVOC concentration. Mean concentrations were calculated for public places, homes and transportation vehicles. Table 1-3 summarizes the mean concentrations of 18 individual VOCs for these locations.

The most visible indoor source of combustion byproducts is ETS. Combustion byproducts of tobacco include CO, particulates, nitrogen

CATEGORY	No. STUDIES	SMOKING			NONSMOKING			DIFF. IN MEANS S − NS
		N	MEAN	RANGE	N	MEAN	RANGE	
OFFICES AND PUBLIC BLDGS.	13	697	2.95	0.1 – 8.7	275	2.99	0.7 – 4.0	−0.04
RESTAURANTS	5	107	3.6	0.4 – 9.0	– –	– –	– –	– –
TAVERNS/BARS	2	5	6.4	– –	– –	– –	– –	– –
TRAINS	2	18	2.2	1.0 – 5.2	10	1.30	0.5 – 2.9	0.90
BUSES	1	35	6.0	3.7 – 10.2	– –	– –	– –	– –
AUTOS	1	– –	– –	– –	213	11.6	8.8 – 22.3	– –
HOMES –	VERY LITTLE INFORMATION ON HOMES; WHERE DATA IS AVAILABLE; 76 HOMES, 0.7 PPM MEAN, RANGE 0 – 4.0 PPM. NO DATA ON SMOKING OR NONSMOKING.							

Figure 1-1. Concentrations of Carbon Dioxide vs. Respirable Suspended Particles, Total Volatile Organic Compounds, Formaldehyde and Nicotine.

Table 1-2. Summary of All Total Volatile Organic Compound Studies Conducted.

LOCATION	No. OF STUDIES	No. OF SITES	AVERAGE ($\mu g/m^3$)	RANGE ($\mu g/m^3$)
PUBLIC PLACES	8	30	315.1	17.85 − 1,627
HOMES	6	174	219.0	
TRANSPORTATION	1	10	149.2	17.35 − 711.5

oxides, aromatic hydrocarbons, acrolein, aldehydes, nicotine, nitrosamines, hydrogen cyanide and ketones[12, 13, 14]. The combination of these byproducts is called ETS. It is a mixture of mainstream smoke exhaled by active smokers, and sidestream smoke which comes off the burning end of a cigarette. When these two mixtures combine in the air they become highly diluted and undergo several chemical changes called aging to form ETS[15]. ETS concentrations in bars, nightclubs, restaurants, lobbies, etc. may be higher especially if ventilation is inadequate.

The two most commonly measured components of ETS are the RSP and nicotine. Even with accurate measurements of RSP in a room, it is estimated that, in most smoking environments, approximately 10% to 50% of the RSP come from ETS. Current estimates of particulate concentrations generated by ETS indoors range from a mean value of 27 $\mu g/m^3$ in homes to in excess of 48 $\mu g/m^3$ in bars and taverns[2]. A review of nicotine concentrations in smoking and nonsmoking areas has been reported[2]. Table 1-4 shows the mean concentrations of nicotine in smoking and nonsmoking homes, offices, restaurants, trains and bars/taverns.

Calculation of dose to ETS has been performed using the mean RSP concentrations from reviewed literature in conjunction with time activity patterns, breathing rates, and percent retention of RSP[2]. Dose is expressed as the mathematical product of concentration, duration, respiration rate and percent retention. Based on these data, it has been shown that, for a number of different exposure scenarios, one's exposure to ETS is typically less than two cigarette equivalents per year and less than four cigarette equivalents per year in the highest exposure scenario. The estimated dose of ETS one can be expected to receive does not support the health risk claims being made by the U.S. Environmental Protection Agency (EPA) and others using epidemiology as a basis.

Table 1-3. Mean Concentrations of 18 Volatile Organic Compounds in Public Places, Homes and Transportation Vehicles.

COMPOUND	PUBLIC PLACES	HOME	TRANSPORTATION
BENZENE S NS	7.4 (9) 7.9 6.5	11.8 (7) 13.5 7.9	26.5 (3) 7.5 29.4
TRICHLOROETHYLENE S NS	25.8 (9) 31.9 18.7	3.3 (6)	
1,1,1-TRICHLOROETHANE S NS	 27.3 60.4	38.4 (8)	14.5 (5)
TETRACHLOROETHYLENE (WITHOUT BAYER DATA) S NS WITHOUT BAYER DATA S WITHOUT BAYER DATA NS	55.8[1] (9) 101.7 26.4	5.2 (8) 2.4 (8) 2.6 2.0	
TOLUENE S NS	40.2 (8) 50.5 29.2	44.1 (5)	87.6 (3)
O-XYLENE S NS	10.5 (9) 14.8 5.7	6.0 (7)	8.5 (3)
M&P XYLENE S NS	32.7 (9) 38.0 26.8	16.0 (6)	82.5[2] (3)
ETHYL BENZENE S NS	7.9 (9) 11.3 4.5	5.7 (4)	6.1 (3)
P-DICHLOROBENZENE	1.1 (4)	8.7 (6)	
STYRENE	8.0 (6)	1.6 (5)	
UNDECANE	16.2 (5)	7.2 (4)	
DECANE	23.9 (6)		
DODECANE	8.3 (5)		
CARBON TETRACHLORIDE		3.2 (5)	
CHLOROBENZENE		4.5 (5)	
OCTANE	18.2 (3)		
CHLOROFORM		4.4 (5)	
A-PINENE	3.8 (4)		

[1] BAYER (1986) DATA – EXTREMELY HIGH VALUES   S = SMOKING
[2] HIGH VALUE DUE TO LOS ANGELES DATA   NS = NONSMOKING
(*) NUMBER OF STUDIES IN PARENTHESES

Table 1-4. Mean Concentrations of Nicotine in Smoking and Nonsmoking Areas for Several Occupancies.

CATEGORY	SMOKING				NONSMOKING			
	No. STUDIES	SAMPLE SITES	MEAN	RANGE	No. STUDIES	SAMPLE SITES	MEAN	RANGE
OFFICES AND PUBLIC BLDGS.	14	673	6.2	ND – 69.7	5	270	0.3	0.1 – 2.1
RESTAURANTS	10	390	5.7	0 – 37.2				
TAVERNS/BARS BETTING SHOPS	4	17	19.1	3 – 65.5	1	2	1.2	0.4 – 2.0
HOMES	1	98	3.7	0.1 – 12.0	3	28	0.29	0 – 1.0
TRAINS	1	20	15.3	0.6 – 49.3	1	20	4.5	0.5 – 21.2

1.3 REFERENCES

[1]Sterling, T.D., and A. Arundel, "Possible Carcinogenic Components of Indoor Air: Combustion Byproducts, Formaldehyde, Mineral fibers, Radiation, and Tobacco Smoke," *J. of Environ. Sci. and Health*, C2(2), pp. 185-230, 1984.

[2]Holcomb, L.C., "Indoor Air Quality and Environmental Tobacco Smoke: Concentration and Exposure," *Environment International*, Vol. 19, pp. 9-40, 1993.

[3]Meckler, M. (Ed), *Indoor Air Quality Design Guidebook*, Ch. 2: Radon, by Elia M. Sterling, The Fairmont Press, Inc./Prentice Hall, 1991.

[4]Steinhausler, F., et al., *J. of Health Physics*, 45, p. 331, 1983.

[5]Meckler, M. (Ed), *Indoor Air Quality Design Guidebook*, Ch. 3: Particulates, by Milton Meckler and Elia M. Sterling, The Fairmont Press, Inc./ Prentice Hall, 1991.

[6]Meckler, M. (Ed), *Indoor Air Quality Design Guidebook*, Ch. 4: Major Combustion Byproducts, by Larry C. Holcomb and Elia M. Sterling, The Fairmont Press, Inc./Prentice Hall, 1991.

[7]Environmental Protection Agency, "Air Quality Criteria for Particulate Matter and Sulfur Oxides," Vol. 1, Environmental Criteria and Assessment Office, Office of Health and Environmental Assessment, NC, 1982.

[8]Turner, S., and P.W.H. Binnie, "An Indoor Air Quality Survey of Twenty-Six Swiss Office Buildings," *Environmental Technology*, Vol.11, pp. 303-314, 1990.

[9]Meckler, M. (Ed), *Indoor Air Quality Design Guidebook*, Ch. 5: Other Indoor Air Pollutants, by Larry C. Holcomb and Elia M. Sterling, The Fairmont Press, Inc./Prentice Hall, 1991.

[10]Holcomb, L.C., "VOC's Reported in Indoor Air Indicate Potential for Avoiding Health and Irritation Problems," *AWMA Annual Meeting*, Denver, CO, 1993.

[11]Holcomb, L.C., and B.S. Seabrook, "Concentration of Volatile Organic Compounds in Indoor Air: Effects on Occupant Comfort and Suggested Controls," in press.

[12]National Academy of Sciences/National Research Council, *Indoor Pollutants*, National Academy Press, Washington, D.C., 1981.

[13]Sterling, T.D., et al., "Indoor By-Product Levels of Tobacco Smoke: A Critical Review of the Literature," *JAPCA*, 32, pp. 250-259, 1982.

[14]Turiel, I., *Indoor Air Quality and Health*, Stanford, CA, Stanford University Press, 1985.

[15]Academy of Science, "Environmental Tobacco Smoke, Measuring Exposures and Assessing Health Effects," National Academy Press, Washington, D.C., 1986.

2

Implications of
ANSI/ASHRAE Standard 62-1989

Milton Meckler, P.E.
President, The Meckler Group
Encino, California

2.1   INTRODUCTION

The high cost of fuels in the 1970s placed immediate pressure on energy conservation. Building construction, maintenance and service practices and standards were revised to allow energy conservation. Buildings constructed in accordance with these energy conservation measures in the last 10 years are well-sealed, mechanically ventilated and centrally controlled, and they have few windows that can be opened. These types of ventilation systems operate with a minimum amount of air. These factors, combined with the emissions from indoor air contamination sources such as synthetic building materials, modern office equipment, and cleaning and biological agents, are believed to increase the indoor air contaminant concentrations.

Control of indoor air quality (IAQ) directly affects the energy consumption of a building due to heating, ventilating and air-conditioning (HVAC) system airflow requirements and air-distribution characteristics. HVAC systems must function to provide an environment where temperature, air distribution, humidity and IAQ are maintained within the allowable levels[1]. The HVAC system designer is responsible to select systems capable of satisfying occupant IAQ needs by maintaining the concentrations of indoor air contaminants as well as energy consumption within acceptable levels. These goals must be accomplished in accordance with ASHRAE Standard 62-1989, which specifies minimum outside airflow requirements for ventilation and acceptable IAQ for human occupancy to avoid adverse health effects. According to the standard, acceptable IAQ involves air in which there are no known contaminants at harmful concentrations as determined by cognizant authorities and with which a substantial majority (80% or more) of the people exposed to it do not express any dissatisfaction.

2.2   ANSI/ASHRAE STANDARD 62-1989

Since the early 1970s, the reduction in the amount of outside air delivered to a building has been considered one of the easiest ways to conserve energy by most building operators, owners and managers. ASHRAE Standard 62-1973 specified minimum and recommended ventilation airflow requirements for most building occupancies; it worked rather well until the energy crisis in the late 1970s. ASHRAE Standard 90-1975 recommended the same outside airflow rates for ventilation as ASHRAE Standard 62-1973. In 1981, ASHRAE Standard 62-1973 was revised to deal with IAQ as well as ventilation odor issues, and it was revised again in 1989.

ASHRAE Standard 62-1981 recommended outside air ventilation rates for smoking-permitted and smoking-prohibited conditions in most spaces. However, the use of this standard for smoking and nonsmoking areas proved to be confusing, and the allowable concentrations of formaldehyde were challenged[2]. Therefore, these rather arbitrary, designated criteria were eliminated from Table 2 of ASHRAE Standard 62-1989, and single outside airflow requirements for ventilation were revised accordingly, taking into account the effect of reasonable smoking. Table 2-1 compares the outside airflow rates for ventilation as required by the Ventilation Rate (VR) Procedure in ASHRAE Standard 62-1989 with those of ASHRAE Standards 62-1973 and -1981.

The outside airflow requirements for ventilation (in cubic feet per minute [cfm] per person) in Table 2 of ASHRAE Standard 62-1989 are for 100 percent outside air when the outside air quality satisfies the specifications for acceptable outside air quality as stated in paragraph 6.1.1 of the standard. While these requirements are for 100 percent outside air flow,

Table 2-1. Comparison of Outside Airflow Rates for Ventilation of ASHRAE Standards 62-1973, -1981 and -1989.

OCCUPANCY	OUTDOOR VENTILATION AIR REQUIREMENT (CFM/PERSON)				
	ASHRAE STANDARD 62-1973		ASHRAE STANDARD 62-1981		ASHRAE STANDARD 62-1989
	MIN.	RECOMM.	NON-SMOKING	SMOKING	VENTILATION RATE PROCEDURE
BALLROMMS, DISCOS	15	20-25	7	35	25
BARS & COCKTAIL LOUNGES	30	35-40	10	50	30
BEAUTY SHOPS	25	30-35	20	35	25
CLASSROOMS	10	10-15	5	25	15
DINING ROOMS	10	15-20	7	35	20
HOSPITAL PATIENT ROOMS	10	15-20	7	35	25
HOTEL CONFERENCE ROOMS	20	25-30	7	35	20
OFFICE CONFERENCE ROOMS	25	30-40	7	35	20
OFFICE SPACE	15	15-25	5	20	20
RESIDENCES	5	7-10	10	10	0.35 ach
RETAIL STORES	7	10-15	5	25	$0.02-0.30$ cfm/ft^2
SMOKING LOUNGES	—	—	—	—	60
SPECTATOR AREAS	20	25-30	7	35	15
THEATER AUDITORIUMS	5	5-10	7	35	15
TRANSPORTING WAITING ROOMS	15	20-25	7	35	15

they also set the amount of air flow required to dilute contaminants to acceptable concentrations. Therefore, it is necessary that at least this amount be delivered to a conditioned space whenever the building is in use, except as modified in paragraph 6.1.3.4 of the standard: *Intermittent and Variable Occupancy.*

In addition to the VR Procedure to achieve acceptable IAQ, ASHRAE Standard 62-1981 also contained an alternative procedure called

IAQ Procedure. This alternative procedure allowed the HVAC designer to use any amount of outside air flow necessary, provided that he could demonstrate the concentrations of indoor air contaminants were below the specified levels. The IAQ Procedure could result in a ventilation airflow rate lower than that of the VR Procedure, but the presence of a particular source of contamination could result in increased ventilation airflow requirements. What the IAQ Procedure failed to account for was the effect that different types of air distribution (constant-volume/variable-temperature vs. constant-temperature/variable-volume) could have on the resulting space contaminant concentration of any given species. This flaw was corrected in the revised ASHRAE Standard 62-1989[3].

ASHRAE Standard 62-1989 now contains a procedure in Appendix E for cleaned, recirculated air that can treat both constant-volume and variable-air-volume (VAV) with constant or proportional outside airflow rates in seven different classes of most widely used HVAC systems. These seven systems and the relationships are listed in Table 2-2[2]. Using the relationships in this table in conjunction with the IAQ Procedure, one can directly compute indoor air contaminant concentrations in a given occupied space, and also verify the adequacy of the outside ventilation airflow rates obtained by the VR Procedure.

One of the major changes in ASHRAE Standard 62-1989 is recognition of the need to maintain a minimum amount of outside air flow per person. Carbon dioxide (CO_2) exhaled by occupants must be diluted to achieve desirable comfort, reduce odors and to avoid serious health hazards. As the need for ventilation load increases, the amount of outside air flow necessary to dilute occupant-generated CO_2 increases. Therefore, CO_2 becomes a convenient surrogate for odor. Five cfm per person as the minimum outside air flow needed, as recommended in ASHRAE Standard 62-1981, has been changed to 15 cfm per person to control occupant odors and ensure that the concentration of CO_2 will not exceed 1000 parts per million (ppm). The rationale for minimum physiological requirements for respiration air based on CO_2 concentration may be found in Appendix D of ASHRAE Standard 62-1989.

One of the most important reasons for documenting the results obtained when employing ASHRAE Standard 62-1989 is to ensure that building operators and managers: (a) understand the design assumptions incorporated into the HVAC system design so that the building can be operated as intended and that when the building is modified or renovated, the HVAC system can be modified accordingly; and (b) maintain

Table 2-2. Required Outside Air Space Contaminant Concentration with Recirculation and Filtration.

CLASS	REQUIRED RECIRCULATION RATE				REQUIRED OUTDOOR AIR	SPACE CONTAMINANT CONCENTRATION	REQUIRED RECIRCULATION RATE
	FILTER LOCATION	FLOW	TEMPER-ATURE	OUTDOOR AIR			
I	NONE	VAV	CONSTANT	100%	$V_o = \dfrac{\dot{N}}{E_v F_r (C_s - C_o)}$	$C_s = C_o + \dfrac{\dot{N}}{E_v F_r V_o}$	NOT APPLICABLE
II	A	CONSTANT	VARIABLE	CONSTANT	$V_o = \dfrac{\dot{N} - E_v RV_r E_f C_s}{E_v (C_s - C_o)}$	$C_s = \dfrac{\dot{N} + E_v V_o C_o}{E_v (V_o + RV_r E_f)}$	$RV_r = \dfrac{\dot{N} + E_v V_o (C_o - C_s)}{E_v E_f C_s}$
III	A	VAV	CONSTANT	CONSTANT	$V_o = \dfrac{\dot{N} - E_v F_r RV_r E_f C_s}{E_v (C_s - C_o)}$	$C_s = \dfrac{\dot{N} + E_v V_o C_o}{E_v (V_o + F_r RV_r E_f)}$	$RV_r = \dfrac{\dot{N} + E_v V_o (C_o - C_s)}{E_v F_r E_f C_s}$
IV	A	VAV	CONSTANT	PROPORTIONAL	$V_o = \dfrac{\dot{N} - E_v F_r RV_r E_f C_s}{E_v F_r (C_s - C_o)}$	$C_s = \dfrac{\dot{N} + E_v F_r V_o C_o}{F_r E_v (V_o + RV_r E_f)}$	$RV_r = \dfrac{\dot{N} + E_v F_r V_o (C_o - C_s)}{E_v F_r E_f C_s}$
V	B	CONSTANT	VARIABLE	CONSTANT	$V_o = \dfrac{\dot{N} - E_v RV_r E_f C_s}{E_v \left[C_s - (1-E_f)(C_o)\right]}$	$C_s = \dfrac{\dot{N} + E_v V_o (1-E_f) C_o}{E_v (V_o + RV_r E_f)}$	$RV_r = \dfrac{\dot{N} + E_v V_o \left[(1-E_f) C_o - C_s\right]}{E_v E_f C_s}$
VI	B	VAV	CONSTANT	CONSTANT	$V_o = \dfrac{\dot{N} - E_v F_r RV_r E_f C_s}{E_v \left[C_s - (1-E_f)(C_o)\right]}$	$C_s = \dfrac{\dot{N} + E_v V_o (1-E_f) C_o}{E_v (V_o + F_r RV_r E_f)}$	$RV_r = \dfrac{\dot{N} + E_v V_o \left[(1-E_f) C_o - C_s\right]}{E_v F_r E_f C_s}$
VII	B	VAV	CONSTANT	PROPORTIONAL	$V_o = \dfrac{\dot{N} - E_v F_r RV_r E_f C_s}{E_v F_r \left[C_s - (1-E_f)(C_o)\right]}$	$C_s = \dfrac{\dot{N} + E_v F_r V_o (1-E_f) C_o}{E_v F_r (V_o + RV_r E_f)}$	$RV_r = \dfrac{\dot{N} + E_v F_r V_o \left[(1-E_f) C_o - C_s\right]}{E_v F_r E_f C_s}$

the building to ensure the safe and proper operation of the HVAC system. The standard contains design documentation procedures requiring the design criteria and assumptions to be documented and made available for operation of the system within a reasonable time following installation[2].

With respect to using the VR Procedure, paragraph 4.2 of the standard states that whenever the VR Procedure is used, the design documentation should clearly state that this procedure was used and, that the design will need to be reevaluated if, at a later time, space-use changes occur or if unusual contaminants or unusually strong sources of specific contaminants are to be introduced into the space. Other documentation requirements may be found in detail in Section 6 and subsections 5.2 and 6.1.3 of the standard.

2.2.1 Outside Air Quality

IAQ is directly related to outside air quality because ventilation using outside air flow is essential to replace the oxygen consumed and to dilute indoor air contaminants to acceptable levels. In fact, this strategy is the basis for achieving acceptable IAQ by the VR Procedure, in which the rate of outside air flow for ventilation to be delivered to an occupied space is specified. However, contaminant concentrations in outside air required by ASHRAE Standard 62-1989 must not exceed the National Primary Ambient Air Quality (NPAAQ) standards for outside air established by the Environmental Protection Agency (EPA), as shown in Table 2-3[2]. Acceptable outside air quality should be evaluated using the definition for acceptable IAQ in Section 3 of the standard.

For example, the NPAAQ standards for outside air require the concentration of particulates to be $50 \, \mu g/m^3$ and $150 \, \mu g/m^3$ for long and short terms, respectively (refer to Table 2-3.) The Air Quality Standards Compliance Report (AQSCR), published monthly by South Coast Air Quality Management District (SCAQMD) in southern California, compares outside air quality by means of readings taken at several designated air-monitoring stations in the South Coast Air Basin and Southeast Desert Air Basin to the state and federal ambient air quality standards.

Since ASHRAE Standard 62-1989 requires that the indoor air contaminant concentrations not exceed the NPAAQ standards in designing within all areas of the SCAQMD, it may be advisable to consider measures that reduce the introduction of particulates known to be potentially harmful to human health[3,4]. The use of air recirculation in conjunction with proper filtration and air-cleaning may provide a more viable alternative

Table 2-3. National Primary Ambient Air Quality Standards for Outside Air Set by Environmental Protection Agency.

CONTAMINANT	LONG TERM			SHORT TERM		
	CONCENTRATION		AVERAGE	CONCENTRATION		AVERAGE
	$\mu g/m^3$	ppm	months	$\mu g/m^3$	ppm	hr
SULFUR DIOXIDE	80.0	0.03	12	365[a]	0.14[a]	24
PARTICLES (PM10)	50.0[b]		12	150[a]	–	24
CARBON MONOXIDE	–	–	–	40,000[a]	35.00[a]	1
CARBON MONOXIDE	–	–	–	10,000[a]	9.00[a]	8
OXIDANTS (OZONE)	–	–	–	235[c]	0.12[c]	1
NITROGEN DIOXIDE	100.0	0.055	12	–	–	–
LEAD	1.5	–	3[d]	–	–	–

[a] NOT TO BE EXCEED MORE THAN ONCE PER YEAR.

[b] ARITHMETIC MEAN.

[c] STANDARD IS ATTAINED WHEN EXPECTED NUMBER OF DAYS PER CALENDAR YEAR WITH MAXIMUM HOURLY AVERAGE CONCENTRATIONS ABOVE 0.12 ppm (235 $\mu g/m^3$) IS EQUAL TO OR LESS THAN 1, AS DETERMINED BY APPENDIX H TO SUBCHAPTER C, 40 CFR 50.

[d] THREE-MONTH PERIOD IS A CALENDAR QUARTER,

(IAQ Procedure) for particulate control in buildings located within the SCAQMD as well as offer the potential to reduce the associated energy cost of having to clean and condition excess (i.e., dilution) outside air. The IAQ Procedure can also be utilized as a check against values established by the VR Procedure to comply with the overall intent of the standard. In so doing, the designer will be able to establish the basis for current and foreseeable building use and set a standard by which building operational data can be periodically checked through the commissioning process[5].

2.3 DYNAMIC MODELING FOR INDOOR AIR QUALITY EVALUATION

As mentioned earlier, Table 2-2 provides relationships to establish design goals and to directly evaluate IAQ for seven classes of HVAC systems. However, these relationships may not provide sufficient information to fully analyze system operation at part-load conditions, and particularly, to predict dynamic variations of indoor air contaminant concentrations throughout the day[6]. Determining maximum concentrations of indoor air contaminants as a function of time during the day can serve as a design strategy to provide IAQ compliance in new construction and remodeling as well as monitoring purposes. It provides a means to control indoor air contaminant concentrations cost-effectively than using excessive outside air (dilution).

In this section, we will first develop a dynamic model for each of the seven most commonly used HVAC systems shown in Table 2-2, and then demonstrate how this dynamic modeling works by providing an example. In this example, we will estimate the concentrations of formaldehyde as a function of time in an office occupancy for three types of ASHRAE-rated filters, and outline how one can choose filters to decrease outside airflow requirement. In addition, we will present the indoor air contaminant concentrations of particulates (PM_{10}) as a function of time for the same office occupancy for monitoring purposes.

Formaldehyde is the most dominant indoor air contaminant in newly constructed and remodeled buildings. A member of a chemical group called aldehydes, formaldehyde is highly reactive and soluble[7]. It may exist as a pure gas, an aqueous solution, and a solid polymer. It is highly irritable to moist body surfaces. Urea-formaldehyde-foam insulation (UFFI), particleboards, some paper products, fertilizers, chemicals, glass and packaging materials are the major sources of formaldehyde. Formaldehyde has been identified as a "carcinogen" and linked particularly to nasal cancer.

2.3.1 Developing a Dynamic Model

Figure 2-1 shows a new model obtained by modifying the model in Appendix E of ASHRAE Standard 62-1989 to include diffusion. Applying a mass-balance for this model gives:

$$m_s = m_g + m_{v,in} - m_{r,out} - m_f - (m_{ia} - m_{ra}) \qquad (2\text{-}1)$$

where

m_g:	mass of contaminant generated in space,
$m_{v,in}$:	mass of contaminant supplied with outside air,
$m_{r,out}$:	mass of contaminant exhausted with return air,
m_f:	mass of contaminant captured by filter,
m_{ia}:	mass of contaminant absorbed by surfaces in space, and
m_{ra}:	mass of contaminant re-absorbed.

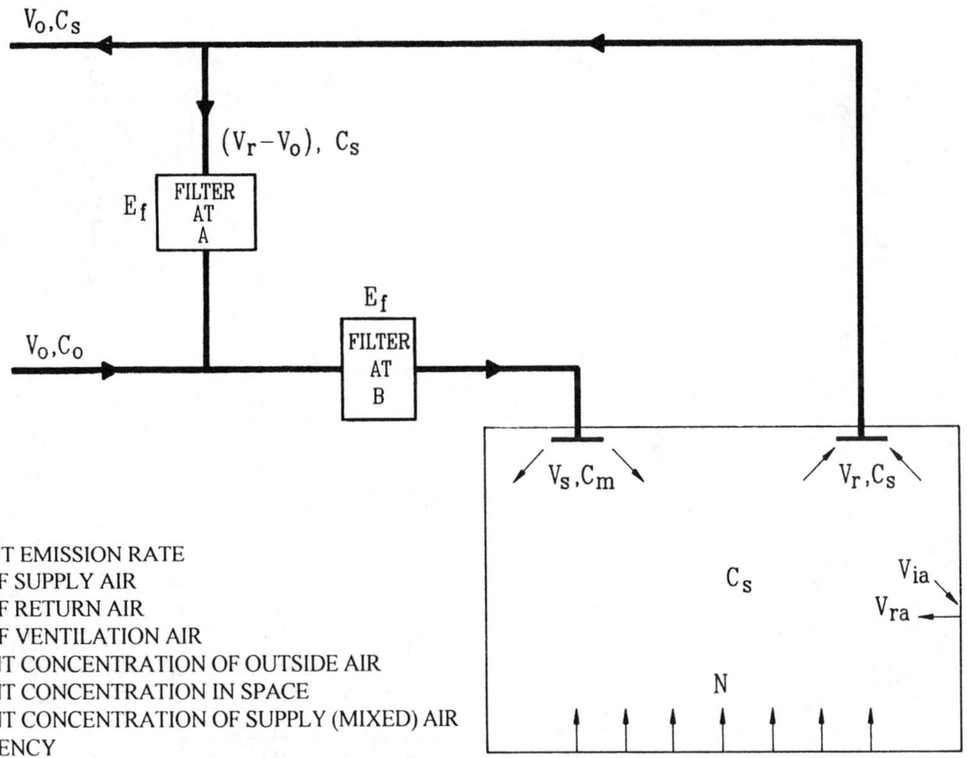

LEGEND

N:	CONTAMINANT EMISSION RATE
V_s:	FLOW RATE OF SUPPLY AIR
V_r:	FLOW RATE OF RETURN AIR
V_o:	FLOW RATE OF VENTILATION AIR
C_o:	CONTAMINANT CONCENTRATION OF OUTSIDE AIR
C_s:	CONTAMINANT CONCENTRATION IN SPACE
C_m:	CONTAMINANT CONCENTRATION OF SUPPLY (MIXED) AIR
E_f:	FILTER EFFICIENCY
V_{ia}:	CONTAMINANT ABSORPTION RATE AND
V_{ra}:	CONTAMINANT RE-ABSORPTION RATE

Figure 2-1. A Modified Model to Determine Concentrations of Contaminants for Filter Locations A and B.

In this model, it is assumed that densities of return air and outside air are the same, contaminant is generated continuously at a steady-rate, and no infiltration or leakage occurs. The filter is either located in the recirculated air (location A) or in the mixed air (location B). Eqn. (2-1) is further simplified by denoting the net effect of absorption ($m_{ia} - m_{ra}$) as m_a, where $m_{ia} > m_{ra}$. The ventilation effectiveness (E_v) is assumed to be 1.0 (perfect mixing). The concentration of a contaminant at any interval of time, dt in a space can be calculated by writing a differential equation for filter location A:

$$QdC_S(t) = Ndt + C_oV_odt - C_S(t)V_odt$$
$$- C_S(t)(V_S - V_O)E_fdt - C_S(t)V_adt \qquad (2\text{-}2)$$

and for filter location B:

$$QdC_S(t) = Ndt + (1-E_f)C_oV_odt$$
$$- C_S(t)V_odt - C_S(t)(V_S - V_O)E_fdt - C_S(t)V_adt \qquad (2\text{-}3)$$

where

$C_S(t)$:concentration of contaminant at time dt,
Q: volume of space,
N: contaminant emission rate,
C_o: concentration of contaminant in outside air,
V_o: flow rate of ventilation air,
V_S: flow rate of supply air,
E_f: filter efficiency, and
V_a: flow rate of absorbed air.

Solving Eqn. (2-2) and Eqn. (2-3) above provides the general solutions in Eqn. (2-4) and Eqn. (2-5) for filter locations A and B, respectively.

$$C_S(t) = C_S(t{-}1) \exp\{-[V_o + V_a + E_f(V_S - V_O)]\,t/Q\}$$
$$+[(C_oV_o + N)/(V_o + V_a + E_f(V_S - V_O))]$$
$$\{1- \exp\{-[V_o + V_a + E_f(V_S - V_O)]\,t/Q\}\} \qquad (2\text{-}4)$$

$$C_S(t) = C_S(t{-}1) \exp\{-[V_o + V_a + E_f(V_S - V_O)]\,t/Q\}$$
$$+\{[(1{-}E_f)C_oV_o + N]/(V_o + V_a + E_f(V_S - V_O))\}$$
$$\{1 - \exp\{-[V_o + V_a + E_f(V_S - V_O)]\,t/Q\}\} \qquad (2\text{-}5)$$

where

$C_S(t{-}1)$: initial concentration of contaminant in space.

Depending on the filter location, either Eqn.(2-4) or Eqn.(2-5) is then solved for $C_S(t)$ for each class of HVAC systems in Table 2-2, therefore, creating a distinct model for each class. The resulting dynamic equations are presented in Table 2-4 for Classes I through VII. Furthermore, the net effect of absorption and re-absorption (or "sink" effects) in Eqn. (2-4) and Eqn. (2-5) is omitted. The contribution of sink effects can be determined by comparing the actually measured contaminant concentrations with the predicted concentrations. Data gathered to date by actual measurements indicate that the sink effects are negligible.

2.3.2 IAQ Evaluation in New Construction and Remodeling

We will now demonstrate how dynamic modeling can be used in estimating the concentrations of formaldehyde in new construction or remodeling. In this example, formaldehyde is assumed to be emitted from resilient flooring, painted surfaces and furniture. The contaminant emission rate of formaldehyde is estimated to be approximately 4.44 $\mu g/m^3$–min per Table H-1 of the draft ASHRAE Standard 62-19XX.

Consider an office occupancy of 1000 ft^2 with a Class VI HVAC system. A maximum occupancy of 7 people per 1000 ft^2 is assumed in accordance with ASHRAE Standard 62-1989. Referring to Table 2-2, the Class VI HVAC system has a VAV system with a filter at location B, and constant temperature and constant outside ventilation airflow rate.

The Class VI HVAC system may have various filter types with different efficiencies. There are two testing methods on which filter efficiency is based. One of these methods that is outlined in ASHRAE Standard 52-76 can be used in evaluating low-, medium- and high-efficiency filters. Although ASHRAE Standard 52-76 was revised in 1993, the new ASHRAE Standard 52.1 does not invalidate its predecessor. The second testing method is the dioctylphthalate (DOP) efficiency testing that evaluates high efficiency particulate air (HEPA) filters.

Figure 2-2 (based on the test method in ASHRAE Standard 52-76) shows the contaminant removal efficiencies of several ASHRAE-rated filters on a mass-mean-diameter (MMD) basis of particulates in microns. For example, the contaminant removal efficiency of an ASHRAE-rated (40%) filter at an MMD of 2.0 microns, is 15%. In our calculations, we will use Type 1 (40%, ASHRAE-rated), Type 2 (60%, ASHRAE-rated) and Type 3 (90%, ASHRAE-rated) filters with corresponding contaminant removal efficiencies of 15%, 50% and 95%.

Table C-1 of the draft ASHRAE Standard 62-19XX provides target

Table 2-4. Contaminant Concentration as a Function of Time for HVAC System Classes I through VII.

HVAC System Class	Filter Location	Flow	Temperature	Outside Air	Space Contaminant Concentration
I	None	VAV	Constant	100%	$C_s(t) = C_s(t-1) + [C_s(t-1) - C_o]\exp(-V_o t/Q) + \dfrac{N}{V_o}[1 - \exp(-V_o t/Q)]$
II	A	Constant	Variable	Constant	$C_s(t) = C_s(t-1)e^{-x} + \dfrac{C_o V_o + N}{V_o + E_f(V_s - V_o)}\left(1 - e^{x}\right)$
III	A	VAV	Constant	Constant	$C_s(t) = C_s(t-1)e^{-y} + \dfrac{C_o V_o + N}{V_o + E_f(F_s V_s - V_o)}\left(1 - e^{y}\right)$
IV	A	VAV	Constant	Proportional	$C_s(t) = C_s(t-1)e^{-z} + \dfrac{C_o F_s V_o + N}{F_s V_o + F E_f(V_s - V_o)}\left(1 - e^{z}\right)$
V	B	Constant	Variable	Constant	$C_s(t) = C_s(t-1)e^{-x} + \dfrac{(1-E_f)C_o V_o + N}{V_o + E_f(V_s - V_o)}\left(1 - e^{x}\right)$
VI	B	VAV	Constant	Constant	$C_s(t) = C_s(t-1)e^{-y} + \dfrac{(1-E_f)C_o V_o + N}{V_o + E_f(F_s V_s - V_o)}\left(1 - e^{y}\right)$
VII	B	VAV	Constant	Proportional	$C_s(t) = C_s(t-1)e^{-z} + \dfrac{(1-E_f)F_s C_o V_o + N}{F_s V_o + E_f F_s(V_s - V_o)}\left(1 - e^{z}\right)$

Note: Exponents x, y and z above are computed as follows:

$$x = \frac{t}{Q}[V_o + E_f(V_s - V_o)] \qquad y = \frac{t}{Q}[V_o + E_f(F_s V_s - V_o)] \qquad z = \frac{tF_s}{Q}[V_o + E_f(V_s - V_o)]$$

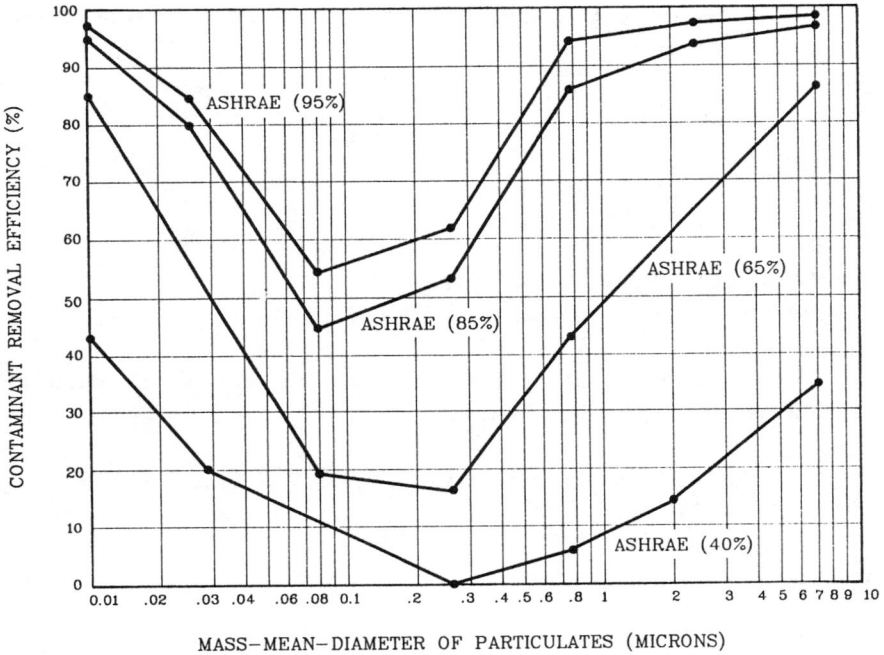

MASS—MEAN—DIAMETER OF PARTICULATES (MICRONS)

Source : EPA Research Triangle

Figure 2-2. Contaminant Removal Efficiency of Several ASHRAE-Rated Filters.

concentration guidelines for most common indoor air contaminants. In using this table, one must realize that contaminants not listed in this table can also cause unacceptable IAQ. For this example, the target (maximum allowable) concentration of formaldehyde is approximately 0.1 ppm (or 122 μg/m^3). Based on this allowable concentration, the calculated outside airflow requirement for the office occupancy is 326 cfm (or 46.5 cfm/ person) for new construction and 178 cfm (or 25.5 cfm/person) for remodeling, if dilution is the only method used to decrease formaldehyde concentrations to allowable levels. These rather high and, therefore costly outside air requirements (in comparison to 20 cfm/person for an office occupancy per ASHRAE Standard 62-1989) may be significantly decreased by the use of air-cleaning in combination with proper filtration. Air-cleaning refers to removal of particulates and both gaseous and vapor phase contaminants.

To demonstrate a sample calculation, the following variables are used:

$E_f = 0.15 \ (15\%)$
$C_o = 0.0 \ \mu g/m^3$
$V_s = 1500 \ cfm$
$V_o = 326 \ cfm$
$Q = 9000 \ ft^3$
$N = 4.44 \ \mu g/m^3$–min (at full occupancy)
F_r = flow reduction factor (refer to Table 2-5)

The C_s of formaldehyde as a function of time can be calculated by solving the following equation in Class VI of Table 2-4.

$$C_s(t) = C_s(t–1) \exp(–y) + \{[(1–E_f) C_o V_o + N]/[V_o + E_f (F_r V_s – V_o)]\} [1 – \exp(–y)]$$

where

$$y = [t/Q] [V_o + E_f (F_r V_s – V_o)]. \tag{2-6}$$

Table 2-5 shows how the C_s of formaldehyde varies hourly (7:00 am to 6:00 pm) depending on the Type 1 filter efficiency for a newly constructed building with variable occupancy. The graph in Figure 2-3 is the graphical presentation of data in Table 2-5 for Type 1 (15%) filter. Similarly, the curves also shown in Figure 2-3 are obtained for Type 2 (50%) filter and Type 3 (95%) filter. For comparison purposes, Figure 2-3 also shows the projected performance with dilution air but without an air-cleaning system. As can be seen from Figure 2-3, filters with higher contaminant removal efficiencies result in considerably decreased indoor air contaminant concentrations in new construction or remodeling.

In summary, dynamic modeling can be used as a design strategy to deal with high concentrations of formaldehyde in new or remodeled buildings. Not only does this strategy verify the compliance of contaminant concentrations obtained by dilution, it also determines the time of day at which maximum concentrations occur. As can be seen from Figure 2-3, maximum formaldehyde concentrations occur between 7:00 am and 9:00 am, and 5:00 pm and 6:00 pm for all three types of filters.

To avoid these maximum concentrations while decreasing the out-

Table 2-5. Concentration of Formaldehyde in an Office Occupancy with a Class VI HVAC System and Type 1 Filter.

TIME OF DAY	FLOW REDUCTION FACTOR (Fr)	OCCUPANCY %	C_s (t) $(\mu g/M^3)$
7:00 am	0.30	5	109.741
8:00	0.40	30	108.939
9:00	0.50	100	103.050
10:00	0.60	100	97.365
11:00	0.65	100	94.575
12:00 noon	0.70	80	92.099
1:00 pm	0.60	40	96.663
2:00	0.80	80	87.867
3:00	0.90	100	83.512
4:00	1.00	100	79.731
5:00	1.00	100	79.598
6:00	0.50	30	100.865

door air requirement to around 20 cfm/person, one needs to use a higher efficiency filter. In this case, holding everything constant, same calculations need to be performed with $V_0 = 20$ cfm/person to observe how these curves behave, and choose the curve with a filter efficiency that will minimize the period of time in which maximum concentrations occur. The

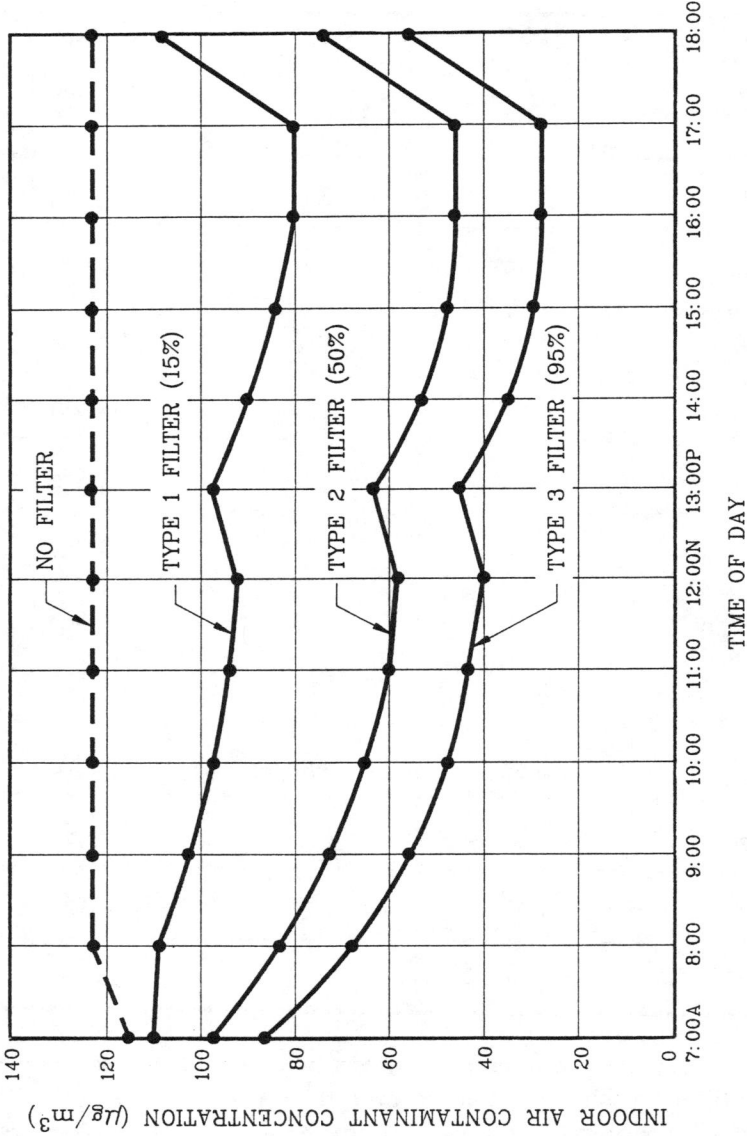

Figure 2-3. Concentration of Formaldehyde in an Office Occupancy with a Class VI HVAC System and Type 1, Type 2 and Type 3 Filters.

dynamic modeling procedure described in this section provides a method to ensure compliance with allowable contaminant concentrations at all times; emphasizes the very important role air-cleaning and filtration play in attaining allowable contaminant concentrations and, therefore acceptable and cost-effective IAQ; and provides a useful means to evaluate HVAC system operation, especially for VAV systems at part-load conditions.

2.3.3   Monitoring Indoor Air Contaminant Concentrations by Dynamic Modeling

Considering the same office occupancy with a Class VI HVAC system as before, let us now estimate $C_S(t)$ of PM_{10} for Type 1 (40%, ASHRAE-rated), Type 2 (60%, ASHRAE-rated) and Type 3 (90%, ASHRAE-rated) filters with corresponding contaminant removal efficiencies of 18%, 56% and 95%. Again, contaminant removal efficiencies of ASHRAE-rated filters in this example are based on an MMD of particulates in microns. The emission rate of PM_{10} is estimated to be approximately $0.018 \, \mu g/m^3$-min per Table H-1 of the draft ASHRAE Standard 62-19XX. Per Table C-1 of the draft ASHRAE Standard 62-19XX, again the maximum allowable C_S of PM_{10} is approximately $50 \, \mu g/m^3$.

The C_S of PM_{10} as a function of time can be calculated by again solving Eqn.(2-6). Figure 2-4 shows how the C_S of PM_{10} varies hourly depending on the Type 1, Type 2 and Type 3 filter efficiencies during the day with variable occupancy. For comparison purposes, Figure 2-4 also shows the projected performance with dilution air but without an air-cleaning system. In this example again (refer to Figure 2-4), filters with higher contaminant removal efficiencies result in considerably decreased C_S of PM_{10}.

As can be seen from Figure 2-4, the monitored maximum PM_{10} concentrations occur between 7:00 am and 9:00 am, and 5:00 pm and 6:00 pm for all three types of filters and they are in compliance with allowable levels. Should these concentrations become significant or exceed the allowable levels, they may be minimized simply by choosing a higher efficiency filter. Choosing an appropriate high-efficiency filter can help outside ventilation airflow rate decrease, resulting in significant energy saving while providing acceptable IAQ.

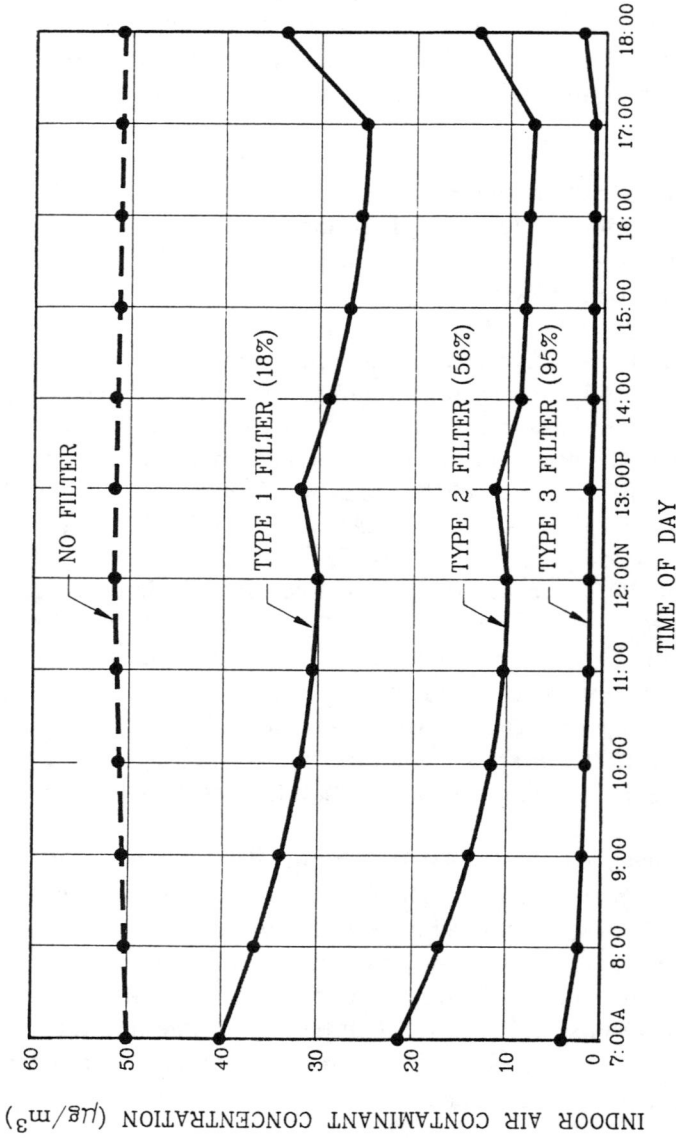

Figure 2-4. Concentration of PM₁₀ in an Office Occupancy with Class VI HVAC System and Type 1, Type 2 and Type 3 Filters.

2.4   REFERENCES

[1]Meckler, M., and J.E. Janssen, "Use of Air Cleaners To Reduce Ventilation," *Proceedings of IAQ '88: Engineering Solutions to Indoor Air Problems*, American Society of Heating, Refrigerating and Air-Conditioning Engineers, Inc., Atlanta, GA., 1988.

[2]ASHRAE Standard 62-1989, *Ventilation for Acceptable Indoor Air Quality*, American Society of Heating, Refrigerating and Air-Conditioning Engineers, Inc., Atlanta, GA, 1989.

[3]Meckler, M., "Ventilation/Air Distribution Cure Sick Buildings," *Specifying Engineer*, 1985.

[4]Swedish Council for Building Research, "Analysis of Low Particulate Size Concentration Levels in Office Environments," *Proceedings of the 3rd International Conference on Indoor Air and Climate*, Vol. 2, Stockholm, Sweden, 1984.

[5]Meckler, M., "Role of Commissioning and Building Operations in Maintaining Acceptable Indoor Air Quality," *Proceedings of Indoor Air '90*, Canada, 1990.

[6]Meckler, M., "Dynamic Response Models for IAQ Performance Evaluation," *Winter Meeting, Seminar 01: Effect of Ventilation Rate and Filtration Efficiency on Indoor Air Contaminants*, American Society of Heating, Refrigerating and Air-Conditioning Engineers, Inc., Atlanta, GA, Jan. 28 - Feb. 1, 1995.

[7]Meckler, M. (Ed), Indoor Air Quality Design Guidebook, Ch. 1: Formaldehyde, Fairmont Press, Inc./Prentice Hall, Lilburn, GA, 1991.

Section 2

Monitoring and Measuring Indoor Air Contaminant Concentrations

3

Development of a Carbon Dioxide Model to Predict IAQ—An Indirect Method

Milton Meckler, P.E.
President, The Meckler Group
Encino, California

3.1 INTRODUCTION

The rate at which oxygen is consumed and carbon dioxide (CO_2) generated depends on the physical activity of the occupants (refer to Figure 3-1). CO_2 exhaled by the occupants in a space must be diluted to achieve desirable comfort as well as to reduce odors and avoid health hazards. CO_2 is now widely recognized as both a convenient surrogate for odor and an indirect measure of the adequacy of supply air (mixing outside and recirculated) to a conditioned space.

A study conducted at a newly constructed federal office building in Portland, Oregon, identified the major sources of volatile organic compounds (VOCs) for long- and short-term source strengths[1]. VOCs generally are used as solvents in the formulation or manufacturing of consumer

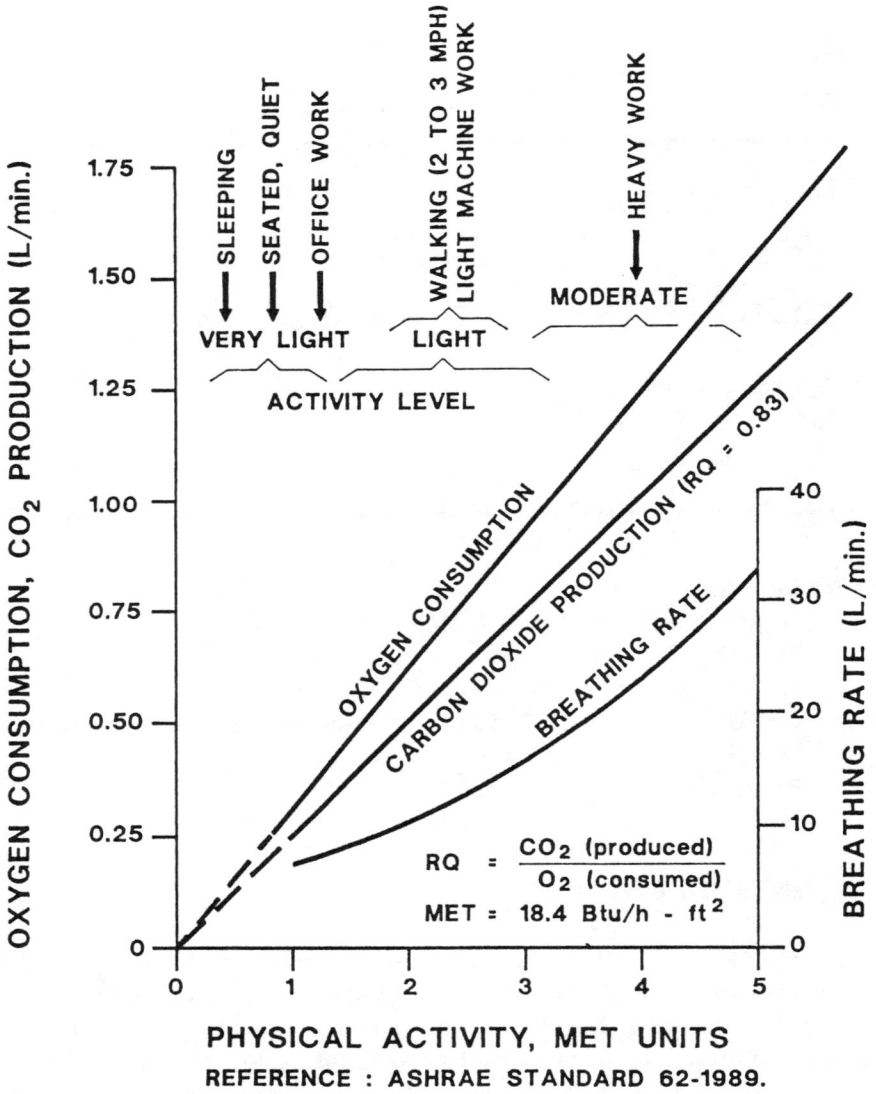

Figure 3-1. Oxygen Consumption and Carbon Dioxide Generation vs. Physical Activity.

products. Table 3-1 shows the types and sources of VOCs commonly found indoors.

The data from the above-referenced study show that the source strengths were substantially higher in mid-morning and mid-afternoon of Friday. The source strengths were again higher the following Monday, with the mid-morning source strengths exceeding the afternoon values. The source strengths were lowest at night and during the weekend, reflecting the low-level of occupancy and activity in the building at these times. The differences between occupied and non-occupied hours demonstrates that the major sources of VOCs were associated with occupant activities.

Although the concentrations of VOCs emitted within a building can be reduced, to some extent, by increasing the ventilation rate, accurately estimated time-varying concentrations of CO_2 can be used as a sensing control strategy to maintain adequate ventilation rates in accordance with ASHRAE Standard 62-1989[2] at all times and all operating conditions. Five cubic feet per minute (cfm) per person, as the minimum outdoor air recommended by ASHRAE Standard 62-1981, has now been substantially increased to 15 cfm per person to control occupant odors and ensure that the concentration of CO_2 will not exceed 1000 parts per million (ppm)[2]. The rationale for minimum physiological requirements for respiration air based on CO_2 concentration may be found in Appendix D of ASHRAE Standard 62-1989.

Table 3-1. Volatile Organic Compounds and Their Sources.

CONTAMINANT TYPE	EXAMPLE	INDOOR SOURCE
ALIPHATIC HYDROCARBONS	PROPANE, BUTANE HEXANE, LIMONENE	COOKING AND HEATING FUELS, AEROSOL PROPELLANTS, CLEANING COMPOUNDS, REFRIGERANTS, LUBRICANTS, FLAVORING AGENTS, PERFUME BASE
HALOGENATED HYDROCARBONS	METHYL CHLOROFORM, METHYLENE CHLORIDE	AEROSOL PROPELLANTS, FUMIGANTS, PESTICIDES, REFRIGERANTS, DEGREASING, DEWAXING AND DRY-CLEANING SOLVENTS
AROMATIC HYDROCARBONS	BENZENE, TOLUENE, XYLENES	PAINTS, VARNISHES, GLUES, ENAMELS, LACQUERS, HOUSEHOLD CLEANERS
ALCOHOLS	ETHANOL, METHANOL	WINDOW CLEANERS, PAINTS, THINNERS, COSMETICS, ADHESIVES, HUMAN BREATH
KETONES	ACETONE	LACQUES, VANISHES, POLISH REMOVERS, ADHESIVES
ALDEHYDES	FORMALDEHYDE, NONANAL	FUNGICIDES, GERMICIDES, DISINFECTANTS, ARTIFICIAL AND PERMANENT-PRESS TEXTILES, UREA-FORMALDEHYDE-FOAM INSULATION (UFFI), PAPER, PARTICLE BOARDS, COSMETICS, FLAVORING AGENTS

REFERENCE : INDOOR AIR QUALITY DESIGN GUIDEBOOK, CHAPTER 5, FAIRMONT PRESS, INC., LILBURN, GA, 1990, EDITED BY MILTON MECKLER.

In this chapter, we will first develop an interactive algorithm capable of estimating time-varying CO_2 concentrations that will be incorporated into a state-of-the-art heating, ventilating and air-conditioning (HVAC) computer load-estimation program. Development of this algorithm is particularly helpful in evaluating the more critical variable-air-volume (VAV) systems that respond to "net space demands" for heating and cooling and still maintain satisfactory balance between indoor air quality (IAQ) and energy consumption. In addition, estimated time- varying CO_2 concentration for a representative Los Angeles high-rise office building, the VAV system of which was recently analyzed (employing various outside design airflow rates ranging from 15 cfm to 20 cfm per person) will be presented.

3.2   A MODEL TO DETERMINE CARBON DIOXIDE CONCENTRATIONS

A mass balance applied to the system shown in Figure 3-2 can be used to determine the concentration of CO_2 in a conditioned space as a function of time assuming equilibrium conditions prevail. A mass balance can be applied as follows:

$$CO_2(t) = f[CO_2 \text{ (produced), Ventilation(in),}$$

$$\text{Ventilation(out), Filtration, Absorption].} \tag{3-1}$$

The following differential equation describes how the concentration of CO_2 varies during the time interval, dt:

$$\begin{aligned} VdC(t) &= Gdt + C_vQ_vdt - C(t)Q_vdt \\ &\quad - C(t)Q_rEdt - C(t)Q_a \, dt \end{aligned} \tag{3-2}$$

and the general solution to Eqn. (3-2) above is given by

$$\begin{aligned} C(t) = \; &C_0\exp[-(Q_v + Q_a + EQ_r)t/V] \\ &+ [(C_vQ_v + G)/(Q_v + Q_a + EQ_r)] \\ &\times \{1 - \exp[-(Q_v + Q_a + EQ_r)t/V]\}. \end{aligned} \tag{3-3}$$

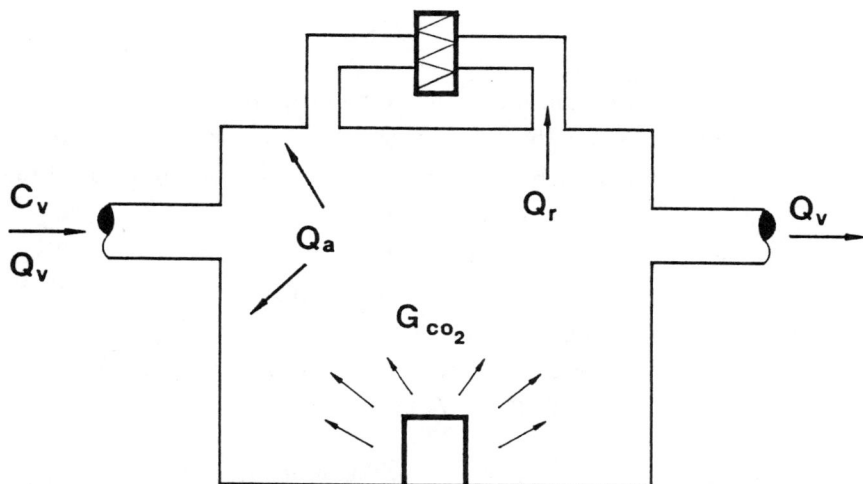

LEGEND

C_v : OUTSIDE AIR CO_2 CONCENTRATION

Q_v : OUTSIDE AIR FLOW RATE

Q_a : ABSORPTION

Q_r : FILTRATION

G_{CO_2} : CO_2 GENERATION RATE PER PERSON

Figure 3-2. A Model to Determine Carbon Dioxide Concentrations.

Assuming the above-referenced absorption, Q_a and filtration EQ_r, terms are both negligible, Eqn. (3-3) simplifies to Eqn. (3-4) below. For brevity, complete derivations leading to Eqn. (3-4) are omitted here. If Eqn. (3-4) is examined carefully, one will see that the second term of the equation decreases with time while the third term increases with time.

$$C(t) = C_v + (C_0 - C_v) \exp(-Q_v t/V)$$
$$+ (G/Q_v) [1 - \exp(-Q_v t/V)] \tag{3-4}$$

where

$C(t)$:   carbon dioxide concentration at time t (ppm);

C_v: outside air carbon dioxide concentration (ppm);

C_0:   initial hourly carbon dioxide concentration (and previous interval's final concentration) (ppm);

Q_v: outside airflow rate, cfm (m^3/h);

V: total volume of conditioned space, ft^3 (m^3);

Q_v/V: air exchange rate, $min^{-1}(h^{-1})$; and

G: carbon dioxide generation rate per person, assumed at 10.59×10^{-3} cfm (18,000 mL/h).

For the variables described above, either the SI or I-P (English) units may be used. The resulting estimated CO_2 concentration will still be expressed in ppm. Also, note that $m^3 = ft^3 \times 0.0283$, if one desires to convert SI units to I-P units.

Table 3-2 is constructed as a sample to show some of the parameters necessary in calculating hourly CO_2 concentrations. These parameters are defined as follows:

a.   "People (%)" is defined as the percentage of the maximum number of occupants on an hourly basis. This profile of space use should be determined by those most familiar with actual building use throughout the day.

Table 3-2. Parameters Used in Calculating Hourly Carbon Dioxide Concentrations.

TIME OF DAY	PEOPLE (%)	MINIMUM OUTSIDE AIR (CFM)	NO. OF PEOPLE PER HR.	OUTSIDE AIR PER PERSON (CFM)	AIR EXCH. RATE (H^{-1})	INITIAL CO_2 CONC. (PPM)	FINAL CO_2 CONC. (PPM)
7 AM	5%	2700	9	300	0.70	400.2	417.9
8	30%	2700	54	50	0.70	417.9	515.6
9	100%	2700	180	15	0.70	515.6	813.3

b. "Minimum outside airflow (cfm)" can be assumed to be constant throughout the day for the "worst case" (maximum) CO_2 concentration scenario. For a given occupancy, Table 2: *Outdoor Air Requirements for Ventilation,* of ASHRAE Standard 62-1989, specifies the minimum outside airflow rate (cfm/person) and the estimated maximum occupancy per 1000 ft^2 of total building area. The specified minimum outside airflow rates are listed for well-mixed airflow conditions (or a ventilation effectiveness approaching unity). For example, for an office occupancy, the minimum outside airflow rate and maximum occupancy are 20 cfm/person and 7 people/1000 ft^2, respectively.

The minimum outside airflow (cfm) equals the outside airflow rate (cfm/person) multiplied by maximum occupancy (number of people/1000 ft^2) and total building area (ft^2).

c. "Number of people per hour" equals the maximum number of people multiplied by "People (%)" in a. above.

d. "Outside airflow per person (cfm/person)" equals the minimum outside airflow (cfm) divided by the actual number of people per hour.

e. "Air exchange rate (h^{-1})" is constant throughout the day. It equals the minimum outside airflow (60 cfm) divided by the volume of conditioned space (ft^3).

f. "Initial carbon dioxide concentration (ppm)" is initially assumed to be 400 ppm and becomes the previous interval's final concentration thereafter.

g. "Final carbon dioxide concentration (ppm)" is calculated for each time period by using Eqn. (3-4) and becomes the initial CO_2 concentration for the following time period.

3.3   AN OFFICE BUILDING STUDY

The study building used to demonstrate the methodology described above has approximately 26,000-ft^2 of floor area and is a high-rise office building in Los Angeles with centrally located chilled water and air-handling units (AHUs) serving VAV terminals located throughout the space. Assume further that there are two AHUs per floor, one serving the perimeter zones and the other serving the interior spaces. In the case of open office areas, the perimeter zone is a floor area up to 15 ft from the exterior building walls. Also assume that heating is to be provided by means of ceiling radiant panels installed on the perimeter zones only. Outside ventilation air is introduced positively by a fan, ducted to the main AHUs through centrally located shafts with intakes located at the roof of the building.

For the purposes of modeling, each floor was assumed to be made up of general office space with approximately 7 people per 1000 ft^2 of occupied space per ASHRAE Standard 62-1989. Since the calculation of CO_2 concentrations is for the worst-case scenario, the minimum outside air supply can be assumed constant throughout the day, regardless of supply air variations. The building occupancy (or space use) profile is based on people occupancy schedules established by the California Energy Commission (CEC) for determining annual energy consumption of high-rise buildings, as shown in Figure 3-3. Consequently, the selected building space has a total number of 180 people and a zone volume of 231,354 ft^3. By using the above-referenced computer program, one can calculate CO_2 concentrations on an hourly basis for design airflow rates of 15 cfm to 20 cfm per person.

Referring the Table 3-3 and Figure 3-4, which show the calculated CO_2 concentrations for a minimum design airflow rate of 20 cfm per person in accordance with ASHRAE Standard 62-1989, note that the CO_2 concentration increases rather sharply between 8:00 am and 11:00 am and then reaches a peak of 904 ppm by 11:00 am. Also, a maximum CO_2 concentration of 921 ppm is reached at 5:00 pm and does not exceed the recommended 1000 ppm. As the space use decreases (due to people leaving the office) between 5:00 pm and 6:00 pm (refer to Figure 3-3), the CO_2 concentration experiences a rather sharp decline and remains constant at around 400 ppm until the next morning.

Figure 3-5 shows that although the generalized shape of the six curves reflects minimum outside ventilation rates (15 cfm to 20 cfm per person), the CO_2 concentrations between 3:00 pm and 5:00 pm remain at or above 1000 ppm. In all six cases, the maximum CO_2 concentration is reached during this time period. As the outside airflow rate is increased from the 15 cfm per person minimum level, the CO_2 concentration tends to decrease. Nonetheless, concentration of CO_2 exceeding 1000 ppm for only a brief period of time during the day does not necessarily indicate unacceptable IAQ but only a transient condition. However, where un-

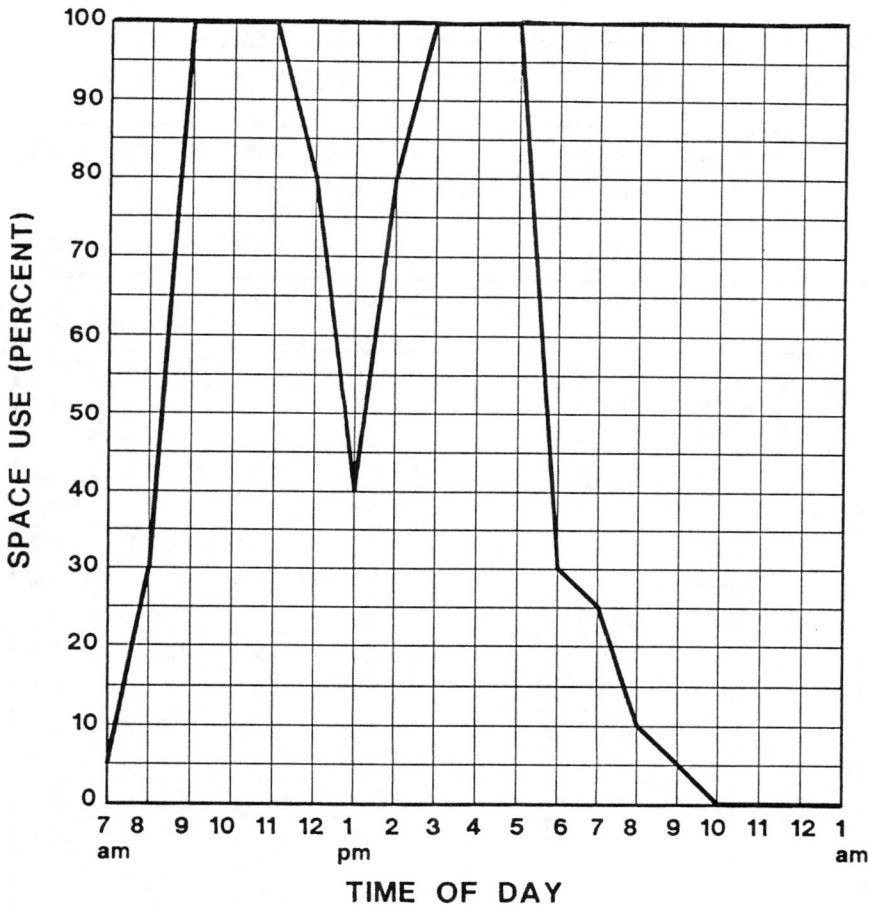

Figure 3-3. Actual Space Use on an Hourly Basis for a Los Angeles High-Rise Office Building.

usual indoor contaminants or sources are present or anticipated, they should be controlled at the source or by the procedure outlined in paragraph 6.2 (IAQ Procedure) of ASHRAE Standard 62-1989.

Table 2 of ASHRAE Standard 62-1989 lists the recommended ventilation airflow rates for several occupancies. In most cases, indoor air contamination is assumed to be directly proportional to the number of people in the space; in other cases, it is assumed to be due mainly to other factors, and the ventilation airflow rates are based on other appropriate parameters.

Table 3-3. Calculated Hourly Carbon Dioxide Concentrations for an Outside Design Airflow Rate of 20 cfm per Person.

TIME OF DAY	PEOPLE (%)	MINIMUM OUTSIDE AIR (CFM)	NO. OF PEOPLE PER HR.	OUTSIDE AIR PER PERSON (CFM)	AIR EXCH. RATE (H^{-1})	INITIAL CO_2 CONC. (PPM)	FINAL CO_2 CONC. (PPM)
7 AM	5%	3600	9	400	0.93	400.0	416.1
8	30%	3600	54	67	0.93	416.1	502.8
9	100%	3600	180	20	0.93	502.8	762.1
10	100%	3600	180	20	0.93	762.1	864.0
11	100%	3600	180	20	0.93	864.0	904.1
12 N	80%	3600	144	25	0.93	904.1	855.5
13 PM	40%	3600	72	50	0.93	855.5	707.7
14	80%	3600	144	25	0.93	707.7	778.3
15	100%	3600	180	20	0.93	778.3	870.4
16	100%	3600	180	20	0.93	870.4	906.6
17	100%	3600	180	20	0.93	906.6	920.8
18	30%	3600	54	67	0.93	920.8	701.2
19	25%	3600	45	80	0.93	701.2	598.8
20	10%	3600	18	200	0.93	598.8	510.3
21	5%	3600	9	400	0.93	510.3	459.5
22	0	3600	0	0	0.93	459.5	423.4
23	0	3600	0	0	0.93	423.4	409.2
24 M	0	3600	0	0	0.93	409.2	403.6
1 AM	0	3600	0	0	0.93	403.6	401.4
2	0	3600	0	0	0.93	401.4	400.6
3	0	3600	0	0	0.93	400.6	400.2
4	0	3600	0	0	0.93	400.2	400.1
5	0	3600	0	0	0.93	400.1	400.0
6	0	3600	0	0	0.93	400.0	400.0

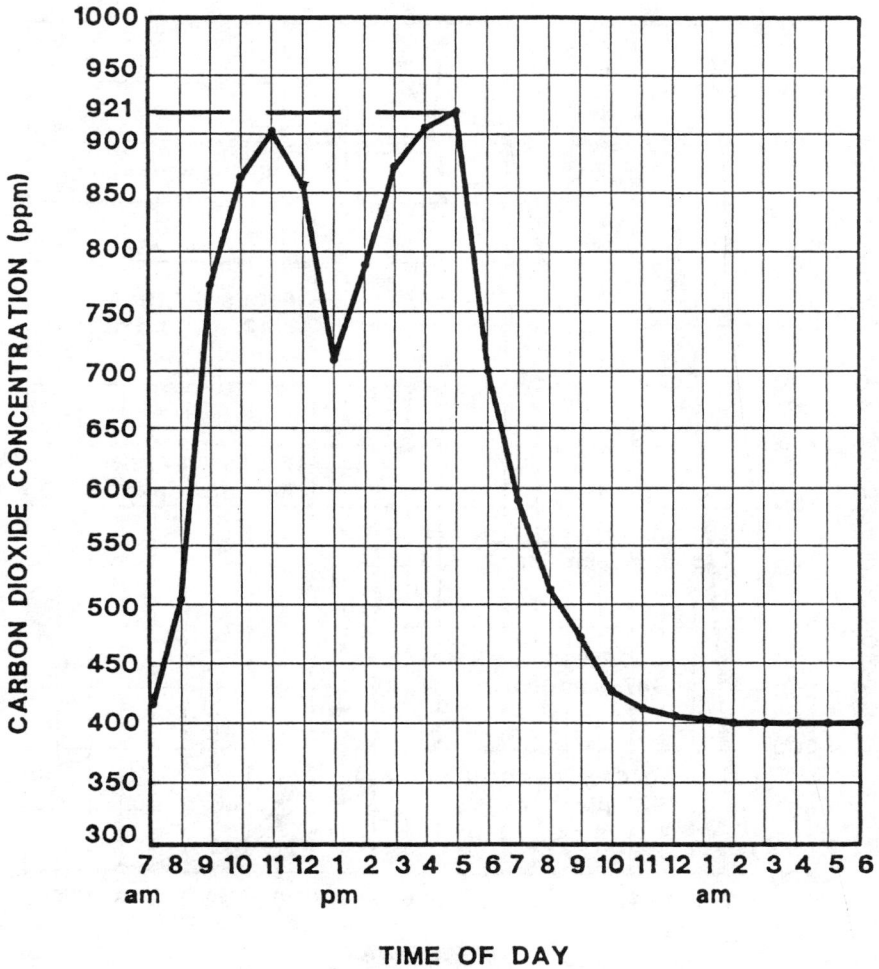

Figure 3-4. Carbon Dioxide Concentration vs. Time of Day for a Minimum Design Airflow Rate of 20 cfm per Person.

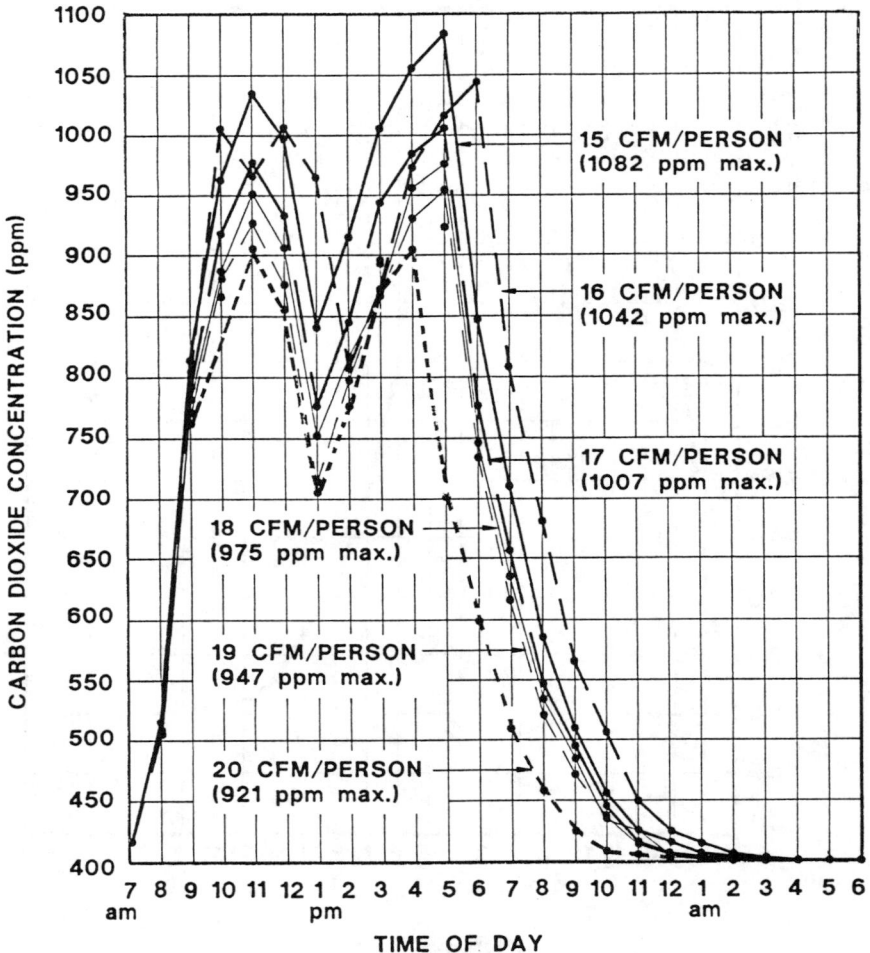

Figure 3-5. Calculated Hourly Carbon Dioxide Concentrations for Minimum Outside Design Airflow Rates of 15 cfm to 20 cfm per Person.

When occupant density differs from what is specified in Table 2 of ASHRAE Standard 62-1989, the "per occupant ventilation airflow rate" for the anticipated occupancy load may be used. The ventilation airflow rates specified in Table 2 of the standard are selected to reflect the consensus that the provision of acceptable outside airflow at these rates will achieve acceptable IAQ by specifically controlling CO_2 concentrations, particulates, odors and/or other indoor air contaminants common to these spaces. Since CO_2 concentration has been widely used as an indicator of IAQ[3], comfort criteria are likely to be satisfied if the ventilation airflow rate is set so that the concentration of CO_2 will not exceed 1000 ppm. If CO_2 is controlled by any method other than dilution, the effects of possible rise in other indoor air contaminant concentrations must be considered.

3.4   MONITORING CARBON DIOXIDE CONCENTRATIONS

Instrumentation to monitor CO_2 concentrations may drift over time. Therefore, it is strongly recommended that the instrumentation be checked periodically and calibrated as needed. CO_2 readings below 300 ppm will not be reliable in most indoor settings since the ambient outside CO_2 concentrations must always exceed inside CO_2 levels (300 ppm to 375 ppm) except during purge cycles when space contaminant concentrations will approach outdoor levels.

Improper location of monitoring instrumentation can also adversely affect the accuracy of CO_2 concentration readings. Even in well-mixed indoor environments, CO_2 gradients will exist around people and furnishings. To avoid either false high or false low CO_2 concentration readings, one should locate the monitoring instrumentation at or near the return air grilles, and operating personnel should be at least six feet away from the instrumentation during calibration. Another factor to avoid when monitoring CO_2 concentrations is the products of combustion. The sources of these combustion products include combustion appliances in or adjacent to the area being monitored, vehicle exhaust at or near parking garages that may infiltrate the building, and re-entrainment of improperly vented combustion fumes from indoors.

One of the advantages of utilizing the CO_2 method is that it can provide reasonable approximate readings to be encountered during full or partial occupancies. Bear in mind that the derivation of the IAQ model assumed a ventilation effectiveness of unity. Therefore, if the air is not uniformly mixed within a breathing zone, higher CO_2 concentrations should be expected.

Properly conducted CO_2 concentration monitoring to ensure measurement accuracy and proper interpretation of results can provide a reliable indication of adequate or inadequate IAQ. However, other factors play an important role in determining acceptable IAQ. For example, presence of VOCs will not normally be detected by a CO_2 sensor. Properly designed CO_2 demand-controlled ventilation (DCV) systems provide an opportunity to maintain acceptable IAQ without excessive and unnecessary use of ventilation air and help reduce associated energy costs. As a practical matter, CO_2 concentrations of 600 ppm to 700 ppm are acceptable since experience shows that IAQ-related complaints above these levels will normally increase.

3.5   USE OF CARBON DIOXIDE CONCENTRATIONS AS A DEMAND-CONTROLLED STRATEGY

Monitoring CO_2 concentrations in a conditioned space can be used as a DCV strategy. A DCV strategy is defined as the effective management of ventilation systems and can be used to maximize IAQ. A DCV strategy based on CO_2 concentrations should only be used in applications where the sources of indoor air contaminants are dominated by occupants of a building. In so doing, a DCV strategy regulates the amount of ventilation in response to changes in building occupancy. During the periods of maximum occupancy, it automatically provides maximum ventilation rates, and during the periods of reduced occupancy, it achieves significant energy savings simply by eliminating excess ventilation.

To illustrate the use of monitoring CO_2 concentrations as a DCV strategy, we will apply the methodology presented in this chapter to the same office building described earlier, and check to ensure that the selected ventilation rates are capable of satisfying minimum ventilation requirements (a "worst-case scenario"). In our study building, outside air is introduced to each equipment room having two AHUs that serve each floor via a separate outside AHU. This AHU is set to deliver outside air either at a constant or variable rate. Assume that commercially available CO_2 sensors are located in the return-air duct to the AHUs and directly interconnected with the building direct digital control (DDC) system.

The three DVC strategies considered are as follows:

- Option 1: Constant outside air flow at 20 cfm per person (in accordance with the VR Procedure in ASHRAE Standard 62-1989) is supplied. The outside AHU was operated during all occupied hours with a two-hour daily purge cycle. Option 1 is used as a base to which Options 2 and 3 are compared.

- Option 2: Required outside airflow rate is controlled by individual CO_2 sensors that are set to maintain a maximum CO_2 concentration of 800 ppm in each conditioned space. The outside AHU was operated only when the building was occupied (6:00 am to 7:00 pm).

- Option 3: Required outside airflow rate is controlled by individual CO_2 sensors that are set to maintain a maximum CO_2 concentration of 920 ppm in each conditioned space. This option closely approximates the conditions in Option 1.

In Figure 3-6, Option 1 is the base case in which outside ventilation airflow rate is constant at 3600 cfm (or 20 cfm/person). Figure 3-6 also shows the outside ventilation airflow rates as a function of time of day for preset maximum CO_2 concentrations of 800 ppm (Option 2) and 920 ppm (Option 3). In Option 2, the calculated maximum outside ventilation airflow rate is 4770 cfm (or 27 cfm/person). In Option 3, the calculated maximum outside ventilation airflow rate is 3672 cfm (or approximately 20 cfm/person). In both Options 2 and 3, the maximums occur between 7:00 am and 11:00 am, and 1:00 pm and 5:00 pm.

An economic analysis[4] compared annual energy savings and associated payback periods of Options 2 and 3 to Option 1 for five representative U.S. cities: Miami; Atlanta; Washington, D.C.; New York; and Chicago. Payback periods were based on the estimated initial cost of equipment and installation of CO_2 sensors. The analysis showed increased energy savings on natural gas for Options 2 and 3 in colder climates. The payback periods for Options 2 and 3 ranged from 1.5 years to 2.0 years. Option 3 proved to be the most favorable one with a lowest payback period for all cities.

The results of this analysis suggest that continuous and accurate monitoring of CO_2 concentrations can be cost-effective in matching the amount of outside air ventilation required for variable occupancy rather than using fixed outside ventilation airflow rates based on peak design occupancy. Such rates are excessive since they fail to account for actual occupancy patterns. Greater levels of comfort as well as significant energy savings can be realized as a result of matching actual occupant ventilation needs on a real-time basis, which is the essence of the DCV strategy.

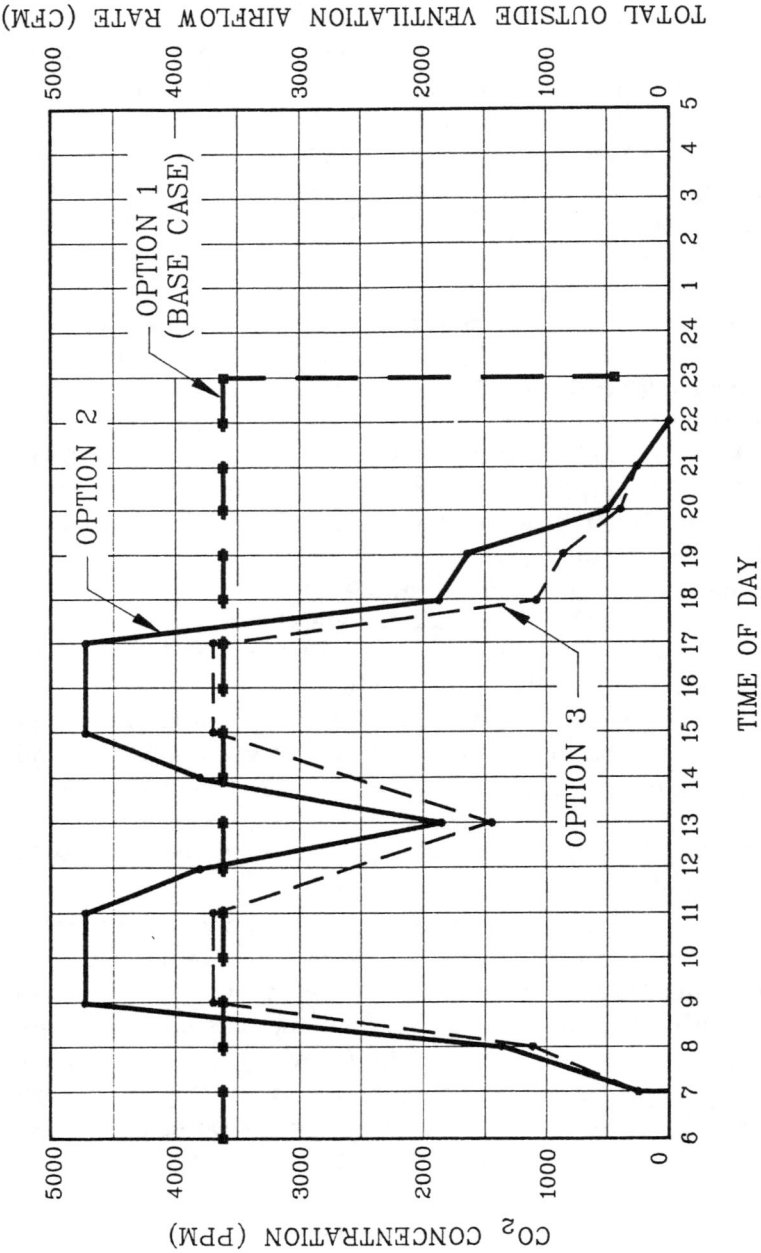

Figure 3-6. Total Calculated Outside Ventilation Airflow Rate as a Function of Time of Day for Options 1, 2, and 3.

3.6   REFERENCES

[1]Hodgson, A.F., et al., "Sources and Source Strengths of Volatile Organic Compounds in a New Office Building," *Journal of the Air Waste Management Association*, Vol. 41, No. 11, pp. 1461-1468, November 1991.

[2]ASHRAE Standard 62-1989, *Ventilation for Acceptable Indoor Air Quality*, American Society of Heating, Refrigerating and Air-Conditioning Engineers, Inc., Atlanta, GA.

[3]Chen, S.Y.S., et al., "Ventilation Analysis for a VAV System," *Heating/Piping/Air Conditioning*, April 1992.

[4]Meckler, M. (Ed), Retrofitting Buildings for Energy Conservation, Ch. 9: Estimating Demand-Controlled Savings on Retrofit Projects, Fairmont Press, Inc., Lilburn, GA, 1994.

4

Development of a Model
To Measure Indoor Air Contaminant
Concentrations — A Direct Method

Milton Meckler, P.E.
President, The Meckler Group
Encino, California

4.1   INTRODUCTION

In addition to the particulates and gaseous air contaminants released during interior building renovations, buildup of combustion byproducts that can come from several different sources such as gas-engine-driven construction equipment can cause serious indoor air quality (IAQ) problems. Among the major combustion byproducts are nitrogen dioxide (NO_2), carbon monoxide (CO), sulfur dioxide (SO_2) and particulates.

Renovation of buildings represents a growing concern for IAQ compliance among the design professionals and building managers and requires careful planning, close supervision and effective communications with people affected by the project. Ventilation requirements originally intended for a given space use may not be adequate if the building is renovated. For example, when an older office space is renovated the effect of different interior finishes, carpets, construction equipment and increased thermal load (due to computers, higher population densities, after-hours use, etc.) can cause an increase in indoor air contaminant concentrations and, therefore in IAQ-related complaints. This may call for rebalancing or adjustment of ventilation airflow rates.

In this chapter, we will first develop a direct method to measure indoor air contaminant concentrations as a function of time in an enclosure. We will then present the results of an actual renovation project in which CO fumes emitted from a gas-engine-driven concrete saw/cutter operated in an unoccupied, renovated-space entered into an adjacent occupied space (a bank) reportedly causing CO-related symptoms. Since the incident was not anticipated, CO concentrations were not measured within the bank and adjacent renovated suite during the operation, and had to be estimated by mathematical modeling as will be described in this chapter. Health effects of CO exposure will also be examined.

4.2   A METHOD TO DETERMINE INDOOR AIR CONTAMINANT CONCENTRATIONS

The concentrations of an indoor air contaminant as a function of time may be determined by using the following first-order differential equation.

$$dC_1/dt = PAC_o + (S/V) - C_1(A + K) \tag{4-1}$$

where

dC_1/dt: rate of increase of indoor air contaminant in time interval dt;

C_1:   instantaneous concentration of indoor air contaminant, mg/m^3;

C_o: initial concentration of indoor air contaminant, mg/m^3;

A: ventilation air changes per hour, ach;

P: penetration factor;

S: emission rate, mg/h;

V: volume, m^3; and

K: absorption factor.

Eqn. (4-1) can be solved by either a conventional method (finding an explicit solution) or numerical integration. A numerical integration method called "backward integration" is utilized here to solve Eqn. (4-1) above. This method uses the previous value, instantaneous slope of the curve, and a time interval, Δt to calculate a new value as shown in Eqn. (4-2) below:

$$x_{t+i} = x_t + \dot{x} \, \Delta t$$

$$(4\text{-}2)$$

where

x_{t+i}:   newly calculated value at time t+i,

x_t: previous value at time Δt

$\dot{x}_t$:   instantaneous slope, and

Δt: time interval.

To demonstrate the use of this method, as an example, in calculating CO concentrations as a function of ventilation and emission rates, Eqn. (4-1) was solved with the following assumptions:

A   =   0.1, 0.15, 0.2 cfm/ft^2;

P   =   1;

C_o =   7.25 mg/m^3;

S   =   3000, 6000, 12,000, 30,000, 60,000 and 90,000 mg/h (for engine sizes of 1/4, 1/2, 1, 2-1/2, 5 and 7-1/2 hp, respectively);

Area   =   3000 ft^2;

V   =   3000 X 8 = 24,000 ft^3 = 679 m^3; and

Δt =   0.05 second.

The CO concentration curves for the above ventilation and emission rates are shown in Figure 4-1. Note that the emission rate is a function of engine size (1/4 hp to 7-1/2 hp). These rates are based on the published information obtained from an engine manufacturer. For brevity, the complete solution of Eqn. (4-1) is omitted here.

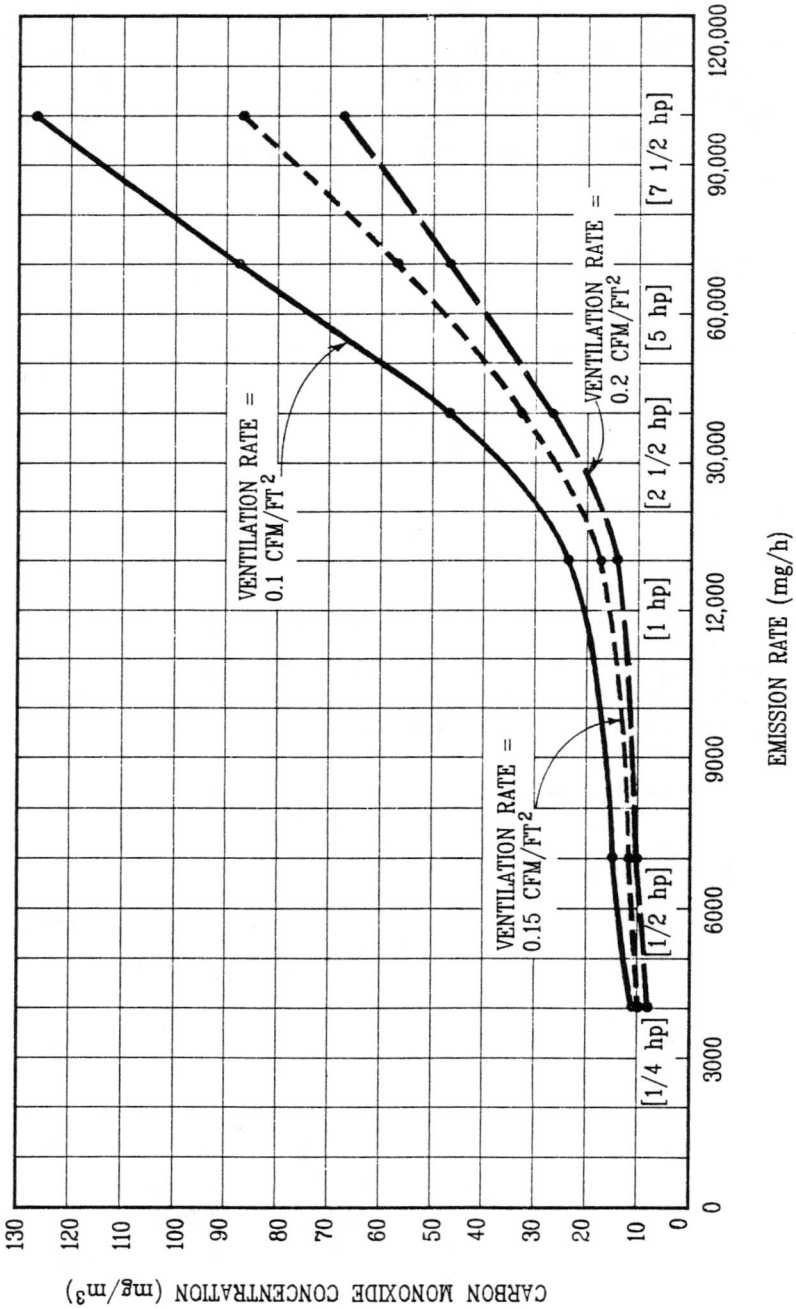

Figure 4-1. Concentration of Carbon Monoxide as a Function of Ventilation and Emission Rates.

4.3   HEALTH EFFECTS OF CARBON MONOXIDE

The toxic effects of CO are due to its combining with the hemoglobin in the blood to form carboxyhemoglobin (COHb). This interferes with the ability of blood to carry oxygen and results in oxygen starvation to body tissues[1]. There is some evidence that suggests CO may also affect the ability of myoglobin and cytochrome oxidase to handle oxygen, thus increasing its ability to deprive tissues of oxygen.

The health effects due to high-level exposure to CO can range from headaches and dizziness to nausea and vomiting and to coma and death. For low-level exposure to CO such as COHb of 2% to 5%, the health effects are not well defined. In young and healthy people, decreased oxygen intake ability and work capacity have been measured at COHb concentrations as low as 5%. Patients suffering from angina have been shown to be affected by COHb levels as low as 2.9%. Neurobehavioral functions have been shown to be affected at COHb levels of 5%. Other health effects due to low-level exposure to CO have been suggested but not confirmed.

Permissible exposure limits (PEL) and threshold limit values (TLV) developed by the Occupational Safety and Health Administration (OSHA) and the American Conference of Governmental Industrial Hygienists (ACGIH), respectively are primarily based on industrial environments. Certain procedures are used for sampling an eight-hour, time-weighted average (TWA) and a 15-minute short-term exposure limit (STEL). Table C-2 of ASHRAE Standard 62-1989 sets guidelines to establish acceptable CO concentration levels for industrial workplaces in the U.S. These guidelines are established as 55 mg/m^3 (50 parts per million [ppm]) on an eight-hour TLV-TWA, and 440 mg/m^3 (400 ppm) on a 15-minute STEL[2].

Since IAQ is a function of various parameters including outside air quality, ASHRAE Standard 62-1989 also recommends that the indoor air contaminant concentrations, as a minimum, should not exceed the limits currently established by the Environmental Protection Agency (EPA) for outside air, collectively known as the National Primary Ambient Air Quality (NPAAQ) standards. These limits are specified in Table 1 of ASHRAE Standard 62-1989, and show short-term average concentrations of 40,000 μg/m^3 (35 ppm) on a one-hour basis and 10,000 μg/m^3 (9 ppm) on an eight-hour basis, for CO.

4.4   A CASE STUDY

A powerful concrete saw/cutter, driven by an air-cooled gas engine, was used during the renovation of a bank building in a southern California desert community. The original bank building consisted of two sections: a bank currently in business, and an adjacent, currently unoccupied, suite that was no longer needed and was to be sublet. The renovation was taking place in this unoccupied suite. Figure 4-2 shows the floor plan of these adjacent spaces along with the actual seating arrangement of bank employees in the reconfigured bank.

During the renovation of the unoccupied adjacent suite, several bank employees in the bank reported what were perceived to be CO-related complaints. To substantiate the validity of attributing a serious illness to the exposure, the complaints were first compiled. Possible symptoms due to other combustion byproducts were investigated and found to be negligible to cause serious health effects. Table 4-1 shows the types of complaints reported by each bank employee and time of their reported complaints. Although there were other people in the bank, their complaints were not taken into consideration in our calculations since they were in constant motion (moving in and out of the bank) and their contribution was negligible. Figure 4-3 shows the percent of increased complaints with respect to the time of day. Note that data in Figure 4-3 and Table 4-1 are for the day when actual concrete cutting process took place.

Referring to Figure 4-3, it can be seen that a gradual increase in complaints occurs between 11:00 am and 2:00 pm as the CO concentration increases, and the maximum number of complaints occur between 2:00 pm and 5:00 pm when the CO concentration reaches a certain limit. At this time, most people exposed to CO already had complaints. Note that Figure 4-3 shows a gradual increase and a sudden drop in complaints. The sudden decrease after 5:00 pm could indicate that most people were already exposed to CO as the CO concentration may have reached its critical level, but considering the bank was closed at 5:00 pm, exposure to CO could not have occurred after that time.

Since the sample size of 10 in this case was too small to conduct any meaningful statistical analysis based on employee complaints, the earlier-mentioned mathematical modeling of CO generation and transfer mechanisms were employed to estimate the buildup of CO concentrations as a function of time during the remodeling of the unoccupied suite as well as the adjacent bank. The results of this simulation were compared to the

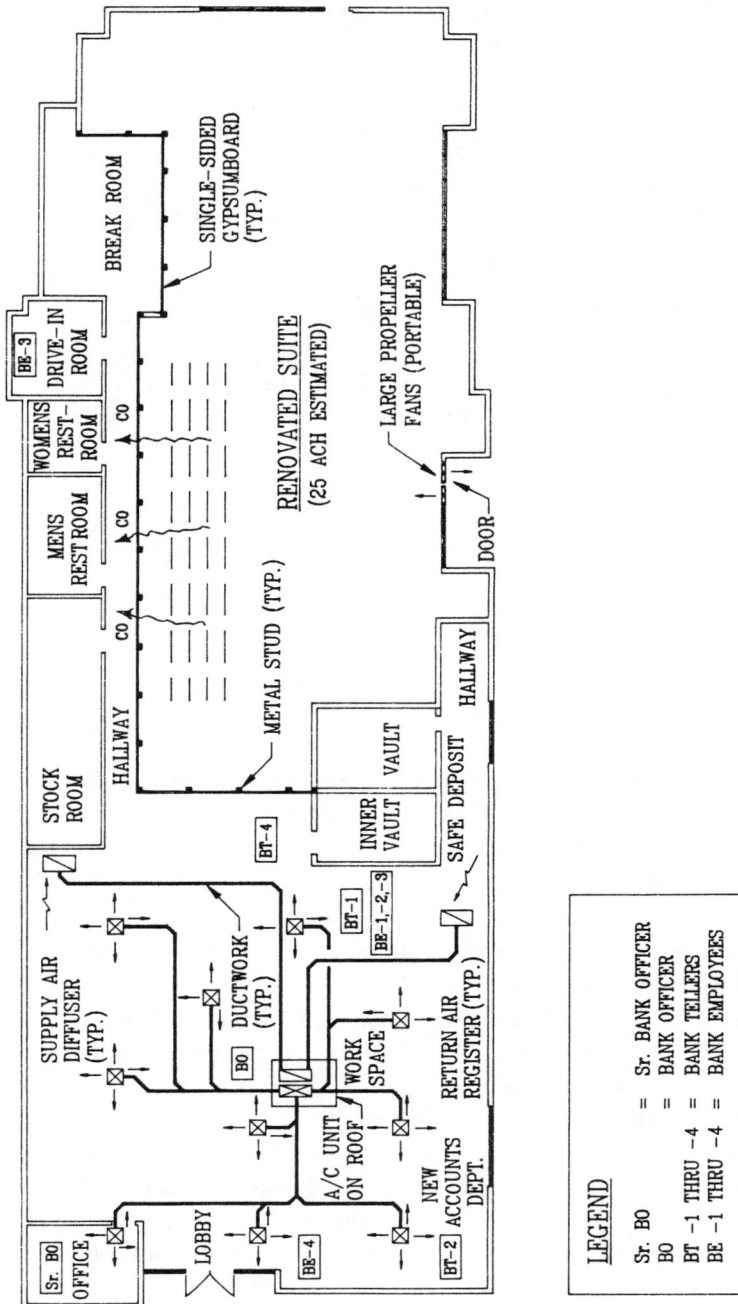

Figure 4-2. Floor Plan of Bank and Adjacent Renovated Suite.

Table 4-1. Types of Complaints Reported by Each Bank Employee and Time of Reported Complaints.

BANK EMPLOYEES	BANK EMPLOYEE DESIGNATION	REPORTED COMPLAINTS BY EACH EMPLOYEE	TIME OF REPORTED COMPLAINTS
Sr. BANK OFFICER	(Sr. BO)	HEADACHE	4:00 PM
BANK OFFICER	(BO)	HEADACHE	2:00 PM
BANK TELLER-1	(BT-1)	SMELLED ODOR	11:00 AM
BANK TELLER-2	(BT-2)	STARTED SMELLING ETHER	MORNING AND 4:00 PM
BANK TELLER-3	(BT-3)	HEADACHE	2:00 PM – 3:00 PM
BANK TELLER-4	(BT-4)	HEADACHE	ALL DAY
BANK EMPLOYEE-1	(BE-1)	HEADACHE AND ON-GOING EYE PROBLEMS	2:00 PM – 4:00 PM
BANK EMPLOYEE-2	(BE-2)	HEADACHE	4:00 PM
BANK EMPLOYEE-3	(BE-3)	SMELLED ODOR	8:30 AM
BANK EMPLOYEE-4	(BE-4)	SMELLED ODOR, EXPERIENCED DIZZINESS, NAUSEA AND ON-GOING HEADACHE	4:00 PM

actually measured COHb levels obtained from a selected group of occupants in the morning of the day following the incident.

Eqn. (4-1) was used to determine the concentrations of CO during the concrete cutting using an emission rate of 18,521 mg/h (for 100 cfm of transfer air) and 32,041 mg/h (for 200 cfm of transfer air) in the bank. The concrete saw/cutter was on for 15 minutes and off for 15 minutes for a total operating period of two hours (or 120 minutes) based on operator statements. Ventilation air exchange rates were estimated to be 25 ach for the renovated suite (based on the exhaust fan placement indicated in Figure 4-2) and 1 ach for the bank.

Figure 4-4 shows the concentrations of CO as a function of time (for two hours during which the cutting process took place) in the renovated unoccupied suite as well as the bank. As can be seen from Figure 4-4, the CO concentration quickly increases to about 210 ppm in the first five minutes of the 15-minute operation period of the concrete cutting, reaches its maximum value approximately at 228 ppm, and experiences a sharp

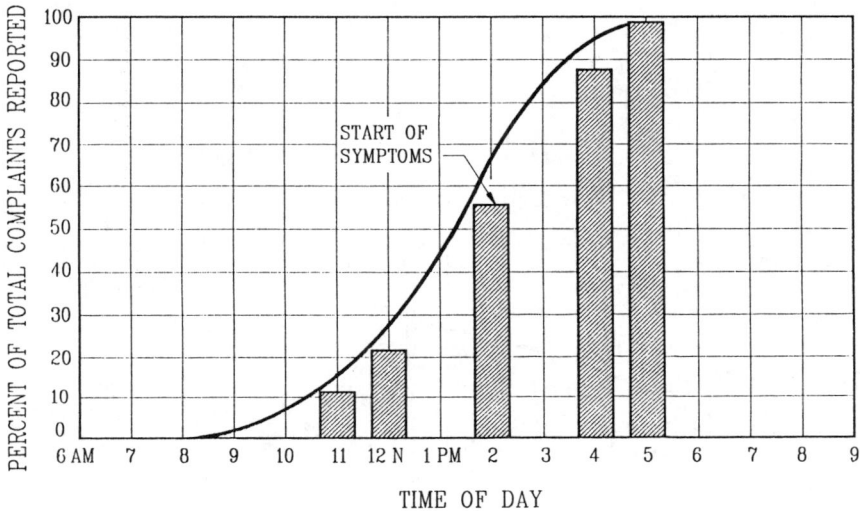

Figure 4-3. Increase of Employee Complaints During Renovation (Percent).

decrease within five minutes immediately following the 15-minute operating period. In the next 10 minutes when the cutting process is off, the CO concentration stayed around 5 ppm. This is a cyclic process that continuous throughout the total operating period of two hours. As shown in Figure 4-4, the CO concentrations in the bank using 100 cfm and 200 cfm transfer air (from the renovated suite to the bank) are about 98 ppm and 100 ppm, respectively at the end of the two-hour period.

4.5   STUDY RESULTS AND CONCLUSIONS

Although the previously mentioned NPAAQ standards in Table 1 of ASHRAE Standard 62-1989 were somewhat exceeded, if Table C-2 of ASHRAE Standard 62-1989 is examined, one can see that the calculated CO concentrations above for the renovated suite and the bank have not exceeded the allowable concentration of 440 mg/m^3 (or 400 ppm) established on a 15-minute STEL[2].

The amount of COHb in the body and, therefore, toxicity of CO depends on the concentration of CO in the air[3,4]. Figure 4-5 shows the percentage of COHb in human blood as a function of exposure time for

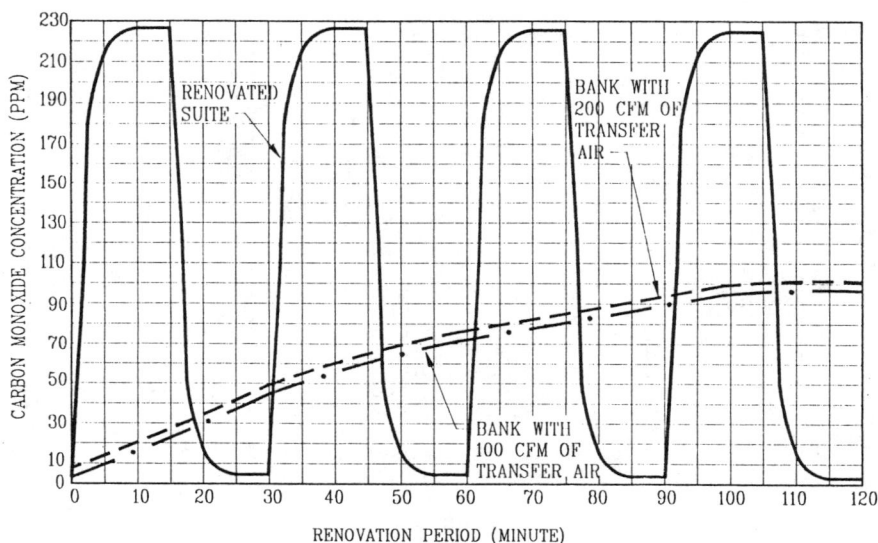

Figure 4-4. Concentrations of Carbon Monoxide as a Function of Time During Renovation for Bank and Adjacent Renovated Suite.

several indoor CO concentrations. If we were to consider the calculated CO concentration of 100 ppm using the maximum estimated 200 cfm transfer air entering the bank, the percent COHb in blood would approach 5% at the end of the two-hour period. The COHb concentration in blood over a period of time is assumed to be cumulative.

According to the earlier-mentioned medical data, the COHb concentrations of 2% to 5% are defined to be low-level exposures. The measured COHb concentrations obtained from the monitored group of the exposed bank employees at a local hospital following the incident were between 4% and 5%. The calculated concentrations of CO and actually measured concentrations of COHb in blood were in general agreement. Therefore, the model and the assumptions made in connection with it, were validated by the actual COHb blood levels obtained from representative employees. Although CO was responsible for the employee complaints and, therefore, IAQ problems reportedly experienced in this case, the calculated concentrations substantiated by COHb blood levels were negligible and unlikely to cause any long term CO-related illnesses, as claimed by the employees.

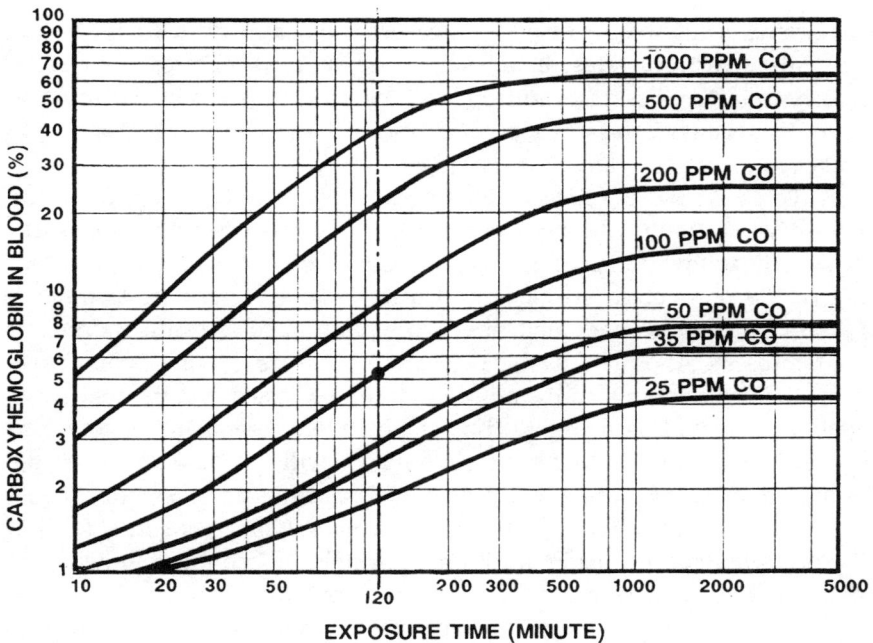

EXPOSURE TIME (MINUTE)

REFERENCE : STEWART, R. D., ET AL. "EXPERIMENTAL HUMAN EXPOSURE
TO CARBON MONOXIDE." ARCH. ENVIRON. HEALTH, 1970.

Figure 4-5. Absorption of Carbon Monoxide in Humans.

4.6   AVOIDING BUILDUP OF
INDOOR AIR CONTAMINANTS DURING RENOVATION

A renovation project requires effective planning. The first step is the renovation in which planning, designing, specifying and relocating take place. Specifying materials, proper choice of mechanical systems, careful review and interpretation of design drawings and operating procedures, retraining of operating and maintenance personnel, and the choice of knowledgeable and suitable contractors and subcontractors are the most important factors in this first step. Second step is the actual renovation process in which demolition, construction, disposal of discarded materials, and commissioning take place. The third step involves the reoccupation of the space and adjustment and rebalancing of heating, ventilating and air-conditioning (HVAC) systems in accordance with the requirements of the renovated space use such as readjustment of the ventilation airflow rates. In all steps, the renovation project must be closely monitored and supervised, and the procedures followed must be in accordance with the safety procedures for both the construction people and occupants of the adjacent spaces.

To avoid the possible IAQ-related problems and, therefore, costly lawsuits associated with them, one must take certain preventive measures during the renovation process. One of these involves the proper partitioning of the renovation area from adjacent spaces as well as sealing off HVAC units to prevent undesired recirculation of harmful indoor air contaminants. Another effective preventive measure is to maintain a pressure differential between the adjacent occupied space and the renovated spaces by providing a slightly higher positive pressure in the occupied space. In this way, indoor air contaminants will not be forced into the occupied spaces. Among other preventive measures are using an exhaust system in the renovated space to outdoors, using 100 percent outside air in the occupied spaces, using additional portable recirculation units in the renovated space and increased amounts of outside air to dilute volatile organic compounds (VOCs), and maintaining outside air ventilation in the renovated space several days following the renovation.

4.7   REFERENCES

[1]Meckler, M. (Ed), *Indoor Air Quality Design Guidebook*, Ch. 4: Major Combustion Products, by Larry C. Holcomb and Elia M. Sterling, Fairmont Press, Inc./Prentice Hall, GA., 1991.

[2]ASHRAE Standard 62-1989: *Ventilation for Acceptable Indoor Air Quality*. American Society of Heating, Refrigerating and Air-Conditioning Engineers, Inc., Atlanta, GA., 1989.

[3]Meckler, M., "Computer Analysis of Smoke Transport During a Hotel Fire," *ASHRAE Transactions*, Vol. 95, Pt. 1, American Society of Heating, Refrigerating and Air-Conditioning Engineers, Inc., Atlanta, GA., 1989.

[4]Paustenbach, D.J., *The Risk Assessment of Environmental Hazards*, Wiley/ Interscience, New York, NY., 1989.

5

Ventilation System Evaluation Using Tracer-Gases

Milton Meckler, P.E.
President, The Meckler Group
Encino, California

5.1   INTRODUCTION

Ventilation is accomplished by natural or mechanical means. Unfortunately, recently developed new building techniques to reduce energy consumption have often resulted in airtight buildings. This has allowed decreased natural ventilation and increased concentrations of indoor air contaminants. The performance of a ventilation system depends on the building geometry, contamination sources, thermal stratification, types and locations of equipment, local mixing of indoor air contaminants, etc. A properly designed ventilation system must achieve a balance between the ventilation rate and contamination sources at all occupied locations for all operating conditions.

In addition to being costly to install, ventilation systems are also costly to operate. When the cost of providing improved ventilation in a building, one must also consider the reduced productivity of personnel and the increased maintenance cost because of inadequate indoor air quality (IAQ). Therefore, it is extremely necessary to monitor a ventilation system closely to ensure that it performs efficiently. Careful monitoring of a ventilation system ensures that occupants of a building receive the required amount of outdoor air in accordance with ASHRAE Standard 62-1989, and that the air is distributed properly in an occupied space.

Ventilation measurements may be made by using tracer-gases. They are usually colorless and odorless inert gases. A tracer-gas selected for monitoring should have a similar density to air and not normally be present in indoor nor outside air. Additionally, its concentration must be measurable to a reasonable accuracy even when it is highly diluted. For safety, comfort and health purposes, the selected tracer-gas must be non-flammable, non-explosive and non-reactive and it should not have any odor nor any adverse health effects in the concentrations used for monitoring. It is also important to ensure that the tracer-gas is not absorbed by the enclosures or the contents of the occupied space and not decompose. Some of the most commonly used tracer-gases are sulfur hexafluoride, nitrous oxide, carbon dioxide and R-12.

There are several system characteristics in evaluating the performance of a ventilation system for acceptable IAQ. In this chapter, we will explore the use of tracer-gases to quantitatively measure two of the major system characteristics: air-exchange rate and adequacy of air-distribution in an occupied space. Since the tracer-gas measurements can be made while the building is occupied, they are more accurate because they take into account the effect of occupancy.

5.2   AIR-EXCHANGE RATE

Air-exchange rate is defined as the ratio of air flow through a space or building to the effective volume of a space or building. Using the continuity equation, as follows:

$$V(dC/dt) = F(t) + Q(t)C_{ir} - Q(t)\,C(t)$$

where

V: volume of air in space (m^3),
C: concentration of tracer-gas in space-air (m^3/m^3),
t: time (h),

F: introduction rate of tracer-gas into space (m^3/h),

C_{ir}:   concentration of tracer-gas into room (m^3/m^3), and

Q: airflow rate through space (m^3/h).

The above differential equation can be rearranged as:

$$Q(t) = [F(t) - V(dC/dt)]/[C(t) - C_{ir}]$$

The air-exchange rate (N) can be found by dividing the air flow rate through the space calculated above by the effective volume of the space. The air flow through a space or building is evaluated using one of the three tracer-gas methods: (a) concentration-decay method, (b) constant-emission method, or (c) constant-concentration method[1].

5.2.1   Concentration-Decay Method

The concentration-decay method is used to obtain discrete air-exchange rates in short periods of time. In this method, a small quantity of tracer-gas is well-mixed with the space air. Upon removal of the source of the tracer-gas, the decay in the concentration of the tracer-gas in the space is monitored over a period of time. To ensure a uniform tracer-gas concentration in a space at any particular time, a large mixing fan is employed during the measurements. If no other tracer-gas is introduced into the space and the air flow through the space during the measurements is constant, the concentration of a tracer-gas decays exponentially as a function of time. If the logarithm of the tracer-gas concentration vs. time is plotted, a straight line is obtained and, the gradient of the line represents the air-exchange rate in the space:

$$N = [\ln C(0) - \ln C(t_1)]/t_1$$

where

N: air-exchange rate (h^{-1}),

C(0):   concentration at time = 0 (m^3/m^3),

$C(t_1)$: concentration at time = t_1 (m^3/m^3), and

t_1:   total measurement period (h).

If the space-air is not well-mixed, a straight line cannot be obtained and the results are invalid.

5.2.2   Constant-Emission Method

The constant-emission method is used for long-term, continuous air-exchange rate measurements in single zones, or for measuring air flow through ventilation ducts. In this method, the tracer-gas is introduced at a constant rate during the measurements. Assuming the air-exchange rate and the tracer-gas concentration in the zone are both constant, the continuity equation is expressed as follows:

$$N = F/(V \times C)$$

Please note that should either the air-exchange rate or introduction rate of the tracer-gas vary during the measurements, the general continuity equation must be used to obtain the air-exchange rate. To ensure a uniform tracer-gas concentration throughout the zone at any time, large mixing fans are employed.

5.2.3   Constant-Concentration Method

The constant concentration method is used for continuous air-exchange rate measurements in one or more zones. In this method, concentrations of a tracer-gas in a zone are measured. Based on this information, a computer controls the amount of a tracer-gas introduced into the zone to keep its concentration constant. A small fan is utilized to mix the tracer-gas with the space-air. In most cases, the air in each zone does not need to be well-mixed. If the concentration of a tracer-gas in a zone is constant over a period of time, the continuity equation is as follows:

$$N(t) = F(t)/(V \times C)$$

The air-exchange rate is directly proportional to the tracer-gas introduction rate required to keep the concentration constant.

This method offers two advantages: it can be used to obtain an accurate long-term, average air-exchange rate in situations where the air-exchange rate varies such as in occupied buildings, and it can also be used to document these variations in detail. The constant-concentration method is also particularly well-suited to the continuous determination of the infiltration of outdoor air into each individual space in a building.

5.3   AIR-DISTRIBUTION IN A SPACE

Air patterns in an occupied space may be as important as the air-exchange rate in that space[2]. Improper air-distribution in a space will impair the effectiveness of the system and require excessive air-exchange rates to compensate, resulting in increased energy consumption and non-uniform mixing. A non-uniform airflow pattern may be due to the temperature of the supply air, temperature gradients within the space, types and positions of supply- and return-air ducts, and the contents of the space. Non-uniform flow can produce localized areas with unusually high concentrations of indoor air contaminants even if the acceptable average concentration for a building is achieved at a given ventilation rate[3].

Proper air-distribution is crucial in all heating, ventilating and air-conditioning (HVAC) applications, especially for variable-air-volume (VAV) systems. The selection of grilles and diffusers plays a major role in proper distribution of air in a conditioned space as they establish the direction and pattern of the air flow. As the fast-moving airstream directed into a conditioned space, it aspirates or induces space air into the moving airstreams. This aspiration helps circulate air in the space and eliminates stratification.

The "throw" is defined as the distance an airstream travels from the outlet to the point of terminal velocity. Terminal velocity is the point at which the discharged air decreases to a given velocity (normally 50 feet per minute [fpm]). Supply air entering a space should be at such a velocity and angle for sufficient mixing. Cool air from a standard diffuser in a VAV system can dump directly down when the volume and velocity are simultaneously reduced, causing inadequate air-distribution within the space. On the other hand, a linear spot diffuser is capable of maintaining a fairly constant throw in a wide range of airflow conditions. The aspirating effect from the moving airstream maintains adequate space air circulation even at reduced airflow rates. The capacity of a slot diffuser is about 50 cfm per foot per slot. VAV systems utilizing blower-powered mixing boxes can employ standard diffusers without reducing air flows.

Because of the aspirating effect, a linear slot diffuser may be set at a lower minimum air quantity than other types of diffusers. The use of no minimum stops results in energy efficiency. When "no-load" conditions exist, the VAV boxes may close completely causing stagnation problems. Minimum stops ensure that each zone has some air movement even during "no-load" conditions. However, reheating the minimum air quantity may result in higher energy consumption. The use of minimum stops can also cause dumping of cold conditioned air.

The Coanda Effect[4] keeps the moving airstream to remain close to adjacent parallel surfaces such as a ceiling or wall. This effect helps to maintain proper air-distribution. The Coanda Effect can be maintained using a flat surface 12 inches to 18 inches away from the outlet of an overhead diffuser. A linear slot diffuser may be placed in perimeter zones in such a way so that the discharge air is perpendicular to the perimeter wall. The velocity of the air striking the wall should be about 150 fpm at full volume. In a heating cycle, any velocity significantly less than 150 fpm will cause a loss of the Coanda Effect along the outside wall. Care should be exercised in placing diffusers.

The common practice of locating both the supply-air outlet and return-air inlet at the ceiling has been shown to result in a ventilation effectiveness of less than 50%. The supply-air and return-air locations should be properly distributed to avoid "dead spaces" in the room. Best mixing is accomplished when the supply-air outlet is high (in the ceiling) and the return-air inlet is low (near the floor) or vice versa.

The age-of-air is an excellent indicator of the distribution of air in a space. The age-of-air is a measure of time the air has been in a space, and it can be defined in two ways: (a) local mean age-of-air, and (b) room average age-of-air. Local mean age-of-air is utilized when the air distribution in naturally ventilated buildings is to be evaluated. It is extremely advantageous because the areas of stagnant air in a space can be located. Room average age-of-air is a quantitative measure of a ventilation system performance. It takes into account both the amount of ventilation air supplied to a space and the distribution efficiency of the air. Room average age-of-air is measured in the return-air duct. This measurement, however, is not reliable where a large proportion of air leaves the space by other means such as through exfiltration.

The air-exchange efficiency plays a very important role on the age-of-air. When there is a perfect piston flow through a space, room average age-of-air is the lowest. If the air in the space is well-mixed, the room average age-of-air will be double the piston flow. The age-of-air in the return-air duct is the same as all other points in the space. If there are areas with stagnant air in a space because of short-circuiting of supply air, the room average age-of-air will be greater than the well-mixed case. To show the advantages of piston flow, consider a space in which both the indoor air contaminants and heat are generated uniformly. In piston flow, return air is hotter and contains a higher indoor air contaminant concentrations than the average in the space. The air-exchange efficiency in a space is defined as the ratio of the local mean age-of-air in the return to twice the room average age-of-air. The local mean age-of-air in the extract is equal to the effective volume of the space divided by the airflow rate. We will now discuss the methods used to determine the age-of-air.

5.3.1 Measurement Methods of Age-of-Air[1]

Three types of tracer-gas methods can be used to measure the age-of-air: (a) pulsed-injection, (b) concentration-growth, and (c) concentration-decay. In the pulsed-injection method, ventilation air entering the space is marked with pulses of a tracer-gas at certain times and the concentrations of the tracer-gas in the return-air duct and at selected points in the space are monitored. Although this method is fast, and requires a comparatively less amount of tracer-gas, it is difficult to obtain rapid enough measurements of tracer-gas concentrations in the space.

In the concentration-growth method, ventilation air is continuously marked with a tracer-gas as it enters the space and the increase in the tracer-gas concentration in the space is monitored. This method is useful where a uniform concentration of a tracer-gas throughout the space is difficult to accomplish. The main disadvantage of this method is that it only measures the supply-air distribution.

In the concentration-decay method, the air in the space is marked with a tracer-gas and the decay of the tracer-gas concentration due to the infiltration of unmarked outside air into the space is monitored. This method is very similar to the concentration-decay method used to measure the air-exchange rate examined earlier except that no space-air mixing takes place after the tracer-gas has been first well-mixed with the space-air. The local mean age-of-air is the area under the curve (tracer-gas vs. concentration time). The measure of the local mean age-of-air at different locations in a space also helps locate the areas of stagnation. The room average age-of-air can also be calculated if a point at which the change in concentration is monitored is in the return-air duct. The air-exchange rate for the entire space can also be calculated. The concentration-decay method is also the only usable method for naturally ventilated spaces.

5.4 VENTILATION SYSTEM MEASUREMENTS

Air flow in ducts can be measured easily and accurately by introducing a tracer-gas at a constant rate and measuring the concentration at downstream of the duct. When using tracer-gas techniques, one must ensure that the distance between the introduction and concentration measuring points is long enough to allow adequate mixing of the tracer-gas across the diameter of the duct. For linear ducts, a distance of 25 times the duct diameter is recommended[1]. In ducts having one or two elbows, a distance of only 10 times the duct diameter will normally be enough because elbows cause turbulence in flow and, therefore better mixing[1]. The mixing at the concentration measuring point can easily be evaluated by taking intermediate concentration measurements at various points across the diameter of the duct. If these intermediate measurements vary a lot, the concentration measurement point should be moved further down the duct. Tracer-gas techniques are also used in the evaluation of short-circuiting of ventilation air.

5.5   ROLE OF TRACER-GAS MEASUREMENT TECHNIQUES IN ASHRAE STANDARDS

The tracer-gas measurement techniques are now under consideration by ASHRAE Standard Project Committee (SPC) 129P. The purpose of the proposed ASHRAE Standard 129P is to devise a method which evaluates the ability of an air-distribution system to provide required amount of ventilation to a zone of a building. This method will include standard techniques that use tracer-gases, and will require the knowledge of data collection, tracer gas analysis, and air volume flow measurements.

The use of this standard may provide the basis for a ventilation effectiveness rating to be used in conjunction with ASHRAE Standard 62-1989. The ventilation effectiveness developed by the standard is referenced to the air delivery performance under well-mixed conditions. A ventilation effectiveness of unity implies air delivery to the measurement location equivalent to that produced by a well-mixed system. A value greater than unity implies air delivery to the measurement location greater than that produced by a well-mixed system. A value less than unity implies air delivery to the measurement location less than that produced by a well-mixed system.

5.6   REFERENCES

[1]Grieve, P.W., *Measuring Ventilation Using Tracer-Gases*, Bruel & Kjaer, Denmark, Oct. 1989.

[2]Fisk, W.J., et al., "Multi-Tracer System for Measuring Ventilation Rates and Ventilation Efficiencies in Large Mechanically Ventilated Buildings," Lawrence Berkeley Laboratory, LBL-20209, 1985.

[3]Janssen, J.E., "Ventilation for Acceptable Indoor Air Quality: Operational Implications," *Proceedings of the International Facilities Management Association Conference*, Houston, TX, 1987.

[4]Meckler, M., and J.E. Janssen, "Use of Air Cleaners to Reduce Outdoor Air Requirements," *Proceedings of the ASHRAE Conference, IAQ '88: Engineering Solutions in Indoor Air Problems*, American Society of Heating, Ventilating and Air-Conditioning Engineers, Inc., Atlanta, GA, 1988.

Section 3

Practical Design Solutions
That Improve
Indoor Air Quality

6

Filtration:
A Cause and Solution for IAQ

H.E. Barney Burroughs
President, IAQ/Building Wellness Consultancy
Alpharetta, Georgia

6.1 INTRODUCTION

Air filtration is emerging as an important element in indoor air quality (IAQ). It is often involved with other factors which create or allow inadequate IAQ. Further, it is a routine recommendation for completing the mitigation process when space must be cleaned up after IAQ-related problems have occurred. Because of this essential aspect of this often misapplied technology, this chapter will discuss air filtration and air-cleaning from the perspective of indoor environment in commercial buildings. The discussion will focus on how both particulate and gas-cleaning equipment work, their background, where they are to be applied, their use to comply with ASHRAE Standard 62-1989, and their contribution to energy management and efficient building operation. First, let us discuss the history of filters.

6.2   HISTORY OF FILTERS

Air filtration is perhaps misunderstood by some and ignored by others. As a result, filtration and its benefit (clean air) is too often overlooked in planning for adequate IAQ. Poor filtration in both residential and commercial buildings has created an entirely new industry called "duct cleaning." Unwanted contaminant buildup (whether in space, ductwork or air-handling unit (AHU) can be a contributor to sick-building-related problems. It can promote microbial growth; expose occupants to respirable particles; and impair the performance of heating, ventilating and air-conditioning (HVAC) system components.

Although the commercial building construction has largely overlooked filtration, American industry has relied heavily on clean air to enable the robust growth of technology since the World War II. Ultra high-efficiency filtration is a relatively old and well-developed technology. It has been over a half century since the development of high efficiency particulate arrestor (HEPA) filters. The importance of this development was that it established 0.3 micron as the most penetrating particle for arrestance-type filters. This has remained as a standard for determining filter performance and has laid the groundwork for 99.97% or virtually absolute filtration efficiency. It also sets the fabrication pattern of deep pleated filter media as the manufacturing style of high-efficiency filters.

HEPA filters were developed during the World War II as part of the Manhattan project to control a particle called radioactive iodine. Iodine has the unique ability of sublimating or jumping from gaseous state to solid state without going through the liquid state. Thus, radioactive iodine appeared around nuclear reactors as a dangerous toxic particle in the form of an airborne condensation nuclei of about 0.3 micron in size. The HEPA filter was developed to protect the environment and the workers from this potential radioactive exposure. The product was later commercialized under patent and marketed under the registered brand name of "absolute" filter. In the 1960s, the technology patents were legally challenged and become public domain to be pursued competitively by other American and international firms.

By the 1960s, hospitals were protecting critical spaces such as operating suites and OB/GYN rooms with extended media bag filters and/or electrostatic precipitators now called electronic air cleaners [EACs]. These filters were used to create relatively pathogen-free supply air in hospitals. The trip to the moon in 1969 was made possible by huge clean rooms where giant computers were built to manage the elaborate flight planning. Later in the space program, special gaseous and particulate filtration units aboard the Lunar Exploratory Module (LEM) allowed the manned exploration of the moon's surface. In the early 1970s, special gaseous filtration media and systems were developed to protect computers and process control apparatus from the gases and contaminants present in hostile industrial settings. Although early systems relied on an activated carbon, this application area was soon dominated by potassium permanganate impregnated alumina. The use of these high-efficiency filtration systems integrated throughout the electronics industry and became the standard for fabrication areas for delicate electronic chips and computer components. These technologies were also used in the pharmaceutical and medical products industries to control corrosive or reactive gases.

There is a long history of highly effective filtration that has been predominantly focused on contaminant control in specialized or industrial spaces. Designers and suppliers of typical HVAC systems in commercial buildings were and still are widely using disposable panel (throwaway) filters that were developed in the 1930s as furnace filters. Furnace filters were developed and used in central hot-air furnaces for the sole purpose of keeping the fire box in the unit from catching fire from flammable accumulations of household lint, hair and other fibers. This type of heating system had gained popularity over a conventional hydronic or radiant-type heating system.

In the 1950s and 1960s, hot-air furnaces and their blowers were essential components of the air conditioners. Furnace filters were cheap, thin (usually no more than one inch in media thickness), had low airflow pressure-drop, and were disposable. Early versions were made of fabric, metal wire screening, or animal hair mats. In later 1930s, they were made of the newly developed man-made spun glass fibers called fiberglass, and the filters were called "duststops." In commercial applications, they were increased to two inches of fibrous glass matt that provided higher dirt holding capacity but did not increase efficiency significantly.

Later developments of low-efficiency (refer to Figure 6-1) panel filters have simply built on the original fibrous matt technology. Fibers were varying to a wide range of lofted natural and synthetic materials. Permanent metal frames and replaceable media pads attained popularity. The media were then mounted on rolls in elaborate machines which automatically advanced as the media loaded. Synthetic polymer media was applied in blankets, or fabricated into wire-reinforced panels using heat sealing techniques. The denier or density of the media was varied and/or combined and various thicknesses were made available. These developments altered and enhanced their dirt-holding, efficiency and airflow characteristics up to a certain degree. This media selection usually was highly influenced by available byproducts or wastes from the fabric and batting manufacturers. Recent modifications have employed the inherent or an imposed electronic charge on the polymeric-type fiber matts to enhance the particle collection and retention properties of filters using such synthetic filter media.

Meanwhile, extended media filters were being developed to serve the needs of specialized spaces. These would include specialized commercial spaces which experience high lint loads such as textile manufacturing plants, retail department stores, marketing fabric products, medical and health-care facilities, and large public assembly buildings. The most compelling marketplace that emerged during this period was the medical and health-care market. In the 1960s and early 1970s, as part of government grant legislation, high-efficiency air cleaning attained wide use in critical care areas such as operating rooms and OB/GYN floors. The electrostatic precipitator air cleaner attained early acceptance for this area of application.

Extended media filter (refer to Figure 6-2) refers to enhancing the surface area of the filtering medium. This was done by pleating or corrugating the media in shallow (and subsequently) in deep pleated pockets or extensions. Other tactics included shaping the media in conical, cylindrical, cubical, wedge or rhomboid forms that increased the surface area of the filter within the given frame size of the filter. The advantage of this extended surface was that the velocity of the air could be reduced, therefore reducing the airflow resistance. This enabled the use of denser filter matts that provided higher efficiency and the retention of smaller size particles. The extended filter surface also yielded higher dirt or holding capacity and resulted in longer useful filter life.

Earlier versions employed elaborate metal wire or mesh frames to support the filter medium that was either porous cellulose or natural fiber matts. As fabrication technologies improved and new filter media were developed, non-supported versions were developed (bag or pocket filters)

a. Throwaway

b. Blankets & Matts

c. Ring Panel

Figure 6-1. Typical Low Efficiency Filters (*Courtesy of Filtration Group*).

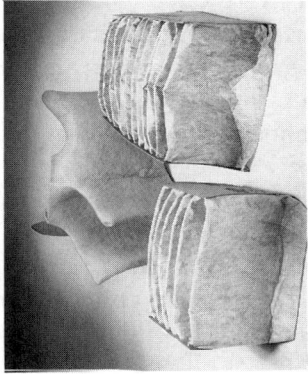

a. Pleated

b.. Cube

c. Supported Basket

Figure 6-2. Typical Extended Media Filter Types (*Courtesy of Filtration Group*).

(refer to Figure 6-3). Highly specialized fabrication techniques resulted in pockets that were sewn with progressive sized stitches, face plates with pockets glued to the metal surface, etc. In a shallow pleated filter, more recent high speed corrugating and fabrication technologies allow the mass production of one-, two- and four-inch pleated filters for medium-efficiency filtration applications.

An early version of the extended media filter developed in the late 1930s employed a porous layered cellulose crepe paper type media. This was pleated into a metal mesh filter retainer system using a manual corrugation machine located on site (PL-24). This led to the prefabricated cartridge which mounted on an elaborate wire support cage that held the filter material rigidly in the airstream (HP). With the development of a superfine fibrous glass filter matt, manufacturers were able to develop filters having much higher efficiency than previously possible.

A more recent development of extended media filters is the combining of pleating, surface extension and HEPA fabrication. This is best represented by a high-efficiency extended media filter which uses corrugation, or "mini-pleating" of HEPA type filter media (refer to Figure 6-4). This is then shaped into a wedge-shaped cartridge that provides even more ultimate surface area. The resulting module provides high efficiency combined with high air flow and long service life.

6.3   APPLYING FILTRATION

Air-cleaning and filtration technologies can be widely used in modern commercial buildings. The following lists the areas of interest to the designer and facility manager:

- Protect mechanical equipment from buildup of energy-consuming films and layers

- Protect occupied space from unsightly dirt splays and dust accumulation

- Protect occupants from irritating or harmful respirable particles

- Protect processes from contamination-caused rejects

- Provide clean make-up air free of external pollutants

a. Rigid Cell   *b. Pocket or Bag*   *c. Minipleat*

Figure 6-3. Typical High Efficiency Filter Types *(Courtesy of Filtration Group).*

a. Conventional   —   *b. Mini Pleat High Capacity*

Figure 6-4. Typical HEPA Type Filters *(Courtesy of Filtration Group).*

- Protect environment from contaminated building exhaust

- Provide source control to spot filter problem space or sources

- Augment ventilation by treating return air for use as equivalent fresh air

The following list outlines some of the potential cost and performance benefits of the appropriate use of filtration.

- Increased system efficiency with maintained high levels of heat exchange efficacy

- Increased system life with reduced wear and contamination

- Lower maintenance costs by avoiding premature cleaning and equipment failures

- Lower housekeeping costs with reduced contaminant loads in a conditioned space

- Avoid product failure due to contamination during the manufacturing process

- Increased productivity of personnel through reduced absenteeism

- Reduced energy consumption because of system cleanliness and heightened efficiency

- Reduced health risk of occupants due to reduction of irritating or pathogenic viable particles

6.4 BASIC PRINCIPLES OF AIR CLEANING

Airborne contaminants are both gaseous and particulate. They vary drastically in size as they are carried in an airstream. Figure 6-5 shows the size of common airborne components. Note that pollen is relatively large with a mean particle diameter of approximately 40 microns to 50 microns.

A micron or micrometer is 1/25,400 of an inch. This can be compared to the even larger human hair that averages around 100 microns in size. Conversely, the size of airborne fungal spores range from 2 microns to 12 microns with a mean particle diameter of approximately 4 to 5 microns. Tobacco smoke is even smaller and is in the submicron range since it is a by-product of combustion and starts out as a condensation nuclei. It is important to understand the size of particulate matter as this is critical to the behavior of filters in controlling that specific size fraction. Gaseous matter is even smaller since these contaminants are carried in solution in an airstream. They are measured in Angstroms equal to one ten-billionth of a meter.

In designing filtration systems, designers should be able to differentiate between efficiency and efficacy. Efficiency is the comparison of incoming quantity to outgoing quantity. For example, a brick wall is a highly efficient filter. Yet, because it will not allow air flow easily, it is not an effective filtering device. Efficacy, on the other hand, includes consideration of the desired result and is a better way of assessing the total performance of an air-cleaning system. To evaluate and to properly specify and apply filtration equipment, a designer must consider other factors besides efficiency. These include factors that influence the overall efficacy, such as:

- Life cycle
- Capacity
- Air flow characteristics
- Efficiency against the contaminant of concern
- Initial cost
- Service life cost/unit time
- Labor of installation and replacement
- Disposal/recycling cost

Life Cycle Cost Assessment (LCCA), widely used in the Green Building movement to express the entire ecological cost impact of a product, is also applicable here. The analysis of the real cost of a filtration system must, therefore, include in addition to its initial cost, its efficacy, labor, life cycle, energy demand, as well as the ecological effect. The latter may take the form of raw material depletion environmental effect from

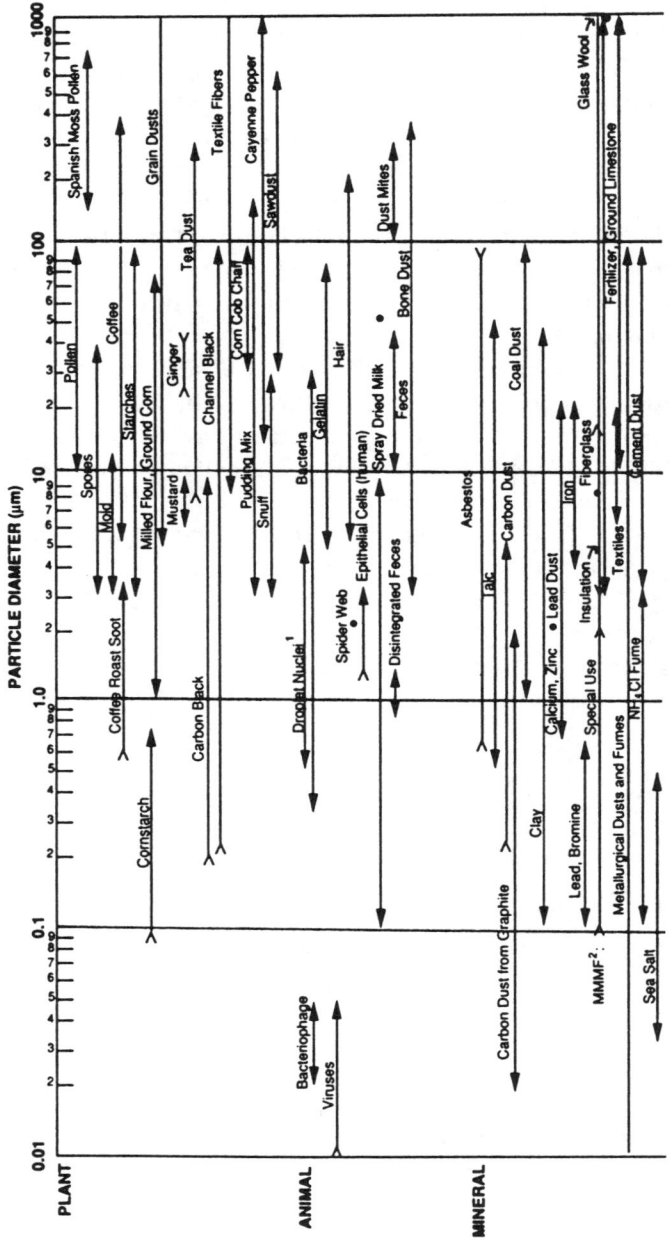

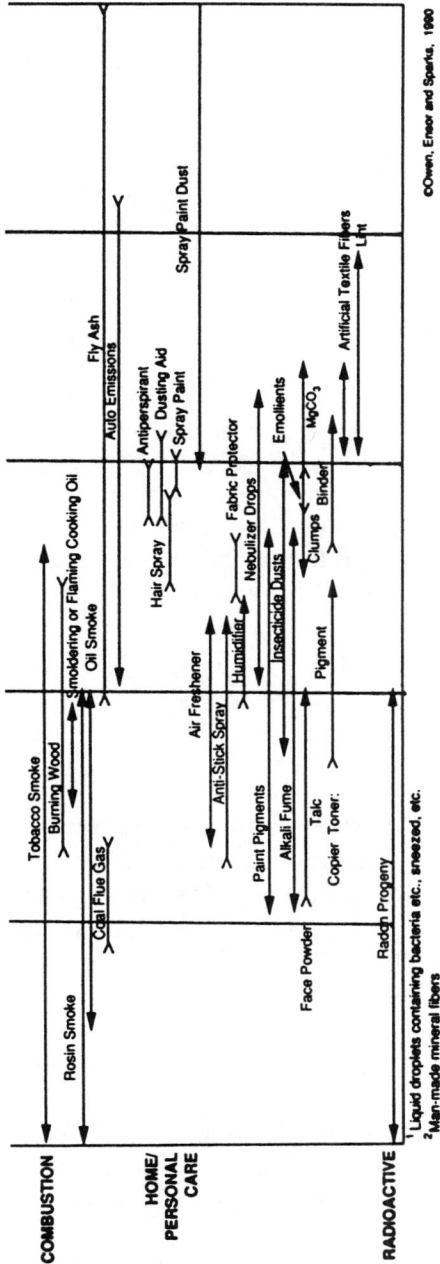

Figure 6-5. Size Distribution of Common Airborne Contaminants.

manufacturing, cost of solid waste disposal and the inherent energy burden. In certain locations where solid waste disposal is a major issue, filters must be completely disassembled into their components to facilitate recycling.

Let us now consider fractional efficiency. Fractional efficiency refers to the efficiency of a particulate filter at specific particle size fractions or against a specific gaseous molecule. For the designer, this addresses more directly the issue of efficacy. For example, what can a designer expect of a filter against a specific component of the airstream such as mold spores or formaldehyde?

6.5   PARTICULATE FILTRATION

We have so far covered both the gaseous and particulate filters. When detailed performance characteristics and application technology are discussed, the specific type of filtration, either gaseous or particulate must be defined. This is because the filtration capture techniques and physical forces vary drastically. The control technology of particulate matter are[1,2]:

* Impingement
* Straining
* Electromagnetic field

Impingement implies a "collision." For example, a particle will impinge (or collide) with a filter fiber and, once intercepted, it will tend to cling to that site (refer to Figure 6-6). This is the predominant mechanism for large particle filtration such as paint booth filtration, as well as the whole range of low-efficiency prefilters. The performance charts will also demonstrate this phenomenon with very small particles less that 0.3 micron in size.

Straining results when the filtration media forms small holes or paths for air to flow through (refer to Figure 6-6). Particles larger than that path will be strained out and then retained within the media. Denser media with smaller air passages will be more effective against smaller particles. Particulate filters relying on either impingement or straining are generically called mechanical filters.

Electromagnetic fields are used to alter and manipulate the electrical

Impingement

Straining

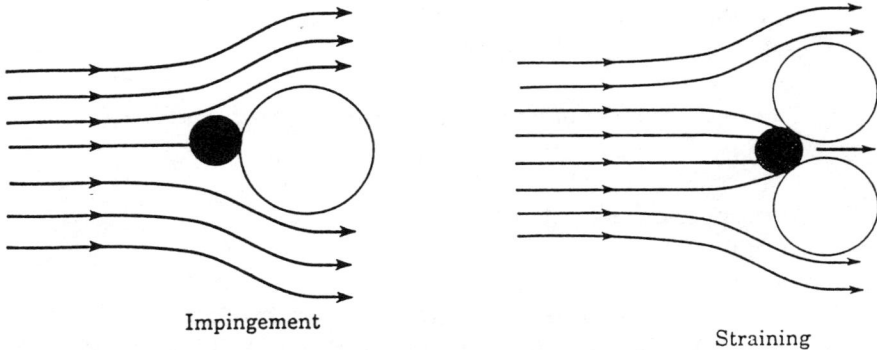

Figure 6-6. Typical Methods of Particle Control.

charge of the airborne particle enabling it to groundout onto a collecting plate or media. The particles are then held into place with electromagnetic forces. Filtering devices relying on an electrical power source and this capture technique are called EAC.

Various media that are used in mechanical filters include paper matt usually in the form of a felt-like matt. Natural fibers such as cotton are also used. Man-made fibers such as superfine fibrous glass and a variety of synthetics are also used in the form of a matt. Certain synthetic fibers have an inherent surface charge or can be altered to enhance their natural charge. In addition to simple straining, this allows the media to employ an electrostatic charge to capture particles. Designers need to be cautioned that electrostatically charged media (referred to as electret) can lose their charge during their life cycle. This causes them to revert into strainer filters which can deteriorate particulate efficiency and retention. This can also be detrimental to their performance and harmful to the airstream.

Efficiency is not always predictable. The factors affecting the particle behavior include its size, mass and charge. In fact, the efficiency of a filter highly depends on particle size. This is why a rating system that does not specify the precise size of the challenge particle may be misleading. Furthermore, the type and loading characteristics of the filter media or device will also affect efficiency and ultimate performance. Strainer media such as that used in bag filters will increase in efficiency as it loads. Thus, a rating system that uses averaging over the life cycle of the filter may provide misleading information on the efficiency of the filter during the early portion of its life. Conversely, EAC filters lose efficiency as the electromagnetic plates are covered with captured contaminants.

6.6   TESTING AND RATING METHODS
FOR PARTICULATE FILTERS

There are a number of filter efficiency testing methods. They include:

- Arrestance Test
- Atmospheric Dustspot Test (ADST)
- DOP Test

The first two methods are covered in American Society of Heating, Refrigerating Air-Conditioning Engineers, Inc. (ASHRAE) Standard 52.1[3]. The Arrestance is the quantity of dust the filter will hold under controlled conditions. It is a gravimetric measurement determined by filter weight gain in comparison to total weight of test dust fed. Thus, it has no relationship to efficiency. This test method is used primarily to differentiate between lower efficiency panel filters. The ADST provides an average efficiency over the life of the filters. It uses atmospheric dust as the challenge and discoloration as the gauge of capture efficiency. This test method is widely used for extended media commercial filters.

The DOP test is based on a military specification which uses Dioctypthalate smoke as the test challenge and a photometer for the determination. Thus, the latter is a fractional efficiency test method and an excellent model. This test method is used to rate ultra high-efficiency filters such as HEPA. Unfortunately, the DOP has been determined to be carcinogenic. Thus, a number of other oils are being used or proposed for an appropriate test aerosol.

ASHRAE Standard 52.1 is in the process of being revised because of inherent flaws. The test methods promulgated in this standard do not address the IAQ needs and are flawed for several reasons. They include:

- ADST uses an undefined and uncontrolled test aerosol (atmospheric air).

- ADST determination is based on discoloration (a 1930s detection technology). Particle counter technology is now available and much more accurate and sensitive.

- ADST data is presented as an average that distorts the actual filter performance and overstates it for much of the filter life.

- Arrestance test yields big and good sounding numbers which are confused with ADST data.

The new *ASHRAE Standard 52.2: Method of Testing General Ventilation Air-Cleaning Devices for Removal Efficiency by Particle Size*[4], was developed and is being promulgated in response to these weaknesses. The test method is based on an ASHRAE-sponsored Research Project 671: *Define a Fractional Efficiency Test Method that is Compatible with Particulate Removal Air Cleaners Used in General Ventilation*.

It is the author's belief that the new test method will solve much of the confusion and misinformation about the performance of general ventilation filters. The primary characteristics of the new test method include the following:

- The test method will evaluate the initial particle removal efficiency of a clean air-filtering device.

- The test aerosol is based on laboratory-generated KCL particles of a defined uniform size range spanning from 0.3 micron to 10 microns.

- The initial clean performance of an air-cleaner is determined as a function of particle size. The performance curve is based on 12 size ranges and is determined by an optical particle counter.

- The test method will evaluate particle removal efficiency over a full-loading life cycle.

The resulting performance data enable a classification according to minimum size fraction efficiency. Table 6-1 shows four categories of performance: coarse (C), low efficiency (L), medium efficiency (M), and high efficiency (H). The data will be reported using the test method in ASHRAE Standard 52.2 as shown in Figure 6-7. This illustration characterizes the efficiency of various types of filters according to 12 bands of particle size (0.3 micron to 10 microns). The data developed using this test methodology can be helpful in determining specific size penetrations of various media types (refer to Figure 6-8).

The testing method presented in the draft ASHRAE Standard 52.2 is applied under controlled laboratory conditions. However, there is recent evidence that the techniques and methods may also be applied in the field.

Table 6-1. Performance Correlation of Filter Efficiency.

Category	Rating	Efficiency at Size Range, μm			Average Arrestance by Std 52.1	Final Resistance ('Minimum) Pa	Final Resistance ('Minimum) in of water
		Range 1 0.30 - 1.0	Range 2 1.0 - 3.0	Range 3 3.0 - 10.0			
Coarse	C1	n/a	n/a	$E_3 < 20\%$	$A_{avg} < 65\%$	150	0.6
	C2	n/a	n/a	$E_3 < 20\%$	$65\% \le A_{avg} < 70\%$	150	0.6
	C3	n/a	n/a	$E_3 < 20\%$	$70\% \le A_{avg} < 75\%$	150	0.6
	C4	n/a	n/a	$E_3 < 20\%$	$75\% \le A_{avg}$	150	0.6
Low Eff.	L5	n/a	n/a	$20\% \le E_3 < 35\%$	n/a	150	0.6
	L6	n/a	n/a	$35\% \le E_3 < 50\%$	n/a	150	0.6
	L7	n/a	n/a	$50\% \le E_3 < 70\%$	n/a	150	0.6
	L8	n/a	n/a	$70\% \le E_3 < 85\%$	n/a	150	0.6
Med. Eff.	M9	n/a	$E_2 < 50\%$	$85\% \le E_3$	n/a	250	1.0
	M10	n/a	$50\% \le E_2 < 65\%$	$85\% \le E_3$	n/a	250	1.0
	M11	n/a	$65\% \le E_2 < 80\%$	$85\% \le E_3$	n/a	250	1.0
	M12	n/a	$80\% \le E_2 < 90\%$	$90\% \le E_3$	n/a	250	1.0
High Eff.	H13	$E_1 < 75\%$	$90\% \le E_2$	$90\% \le E_3$	n/a	350	1.4
	H14	$75\% \le E_1 < 85\%$	$90\% \le E_2$	$90\% \le E_3$	n/a	350	1.4
	H15	$85\% \le E_1 < 95\%$	$90\% \le E_2$	$90\% \le E_3$	n/a	350	1.4
	H16	$95\% \le E_1$	$95\% \le E_2$	$95\% \le E_3$	n/a	350	1.4

n/a   - Not Applicable
A_{avg} - Average Arrestance by Standard 52.1

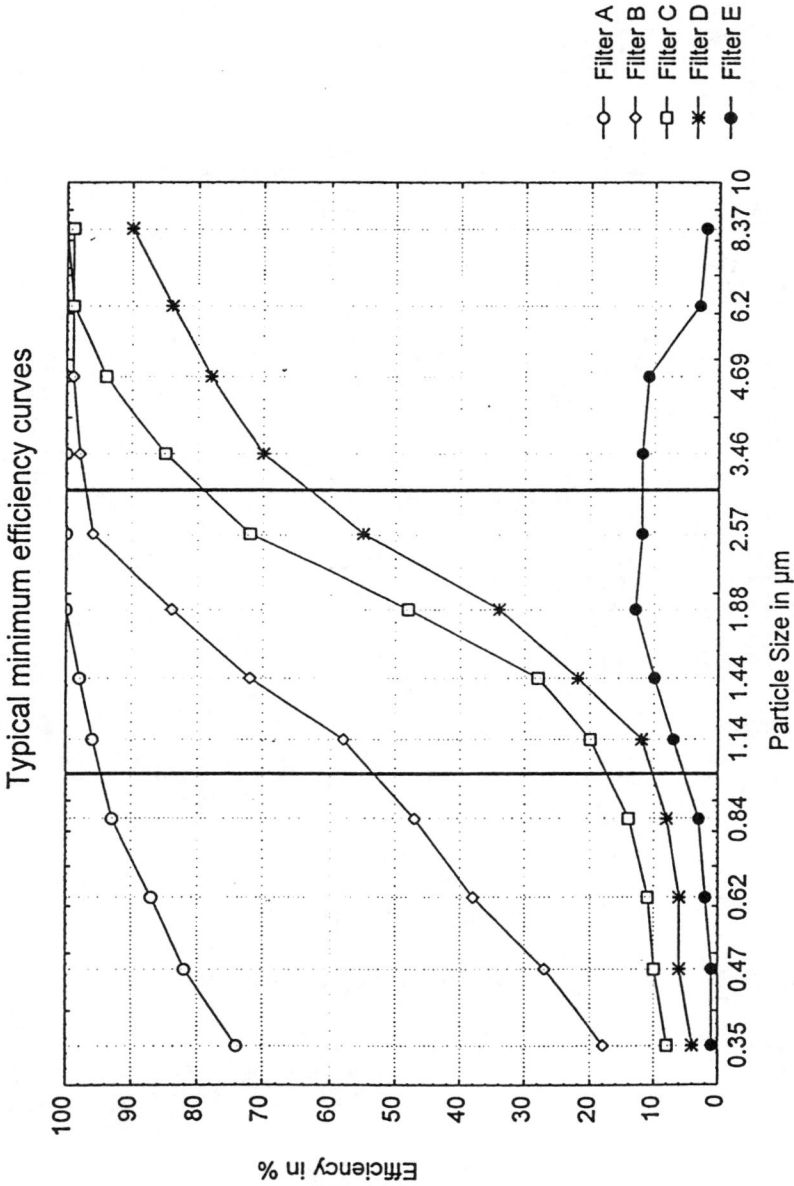

Figure 6-7. Filter Performance Curves of Typical Filters.

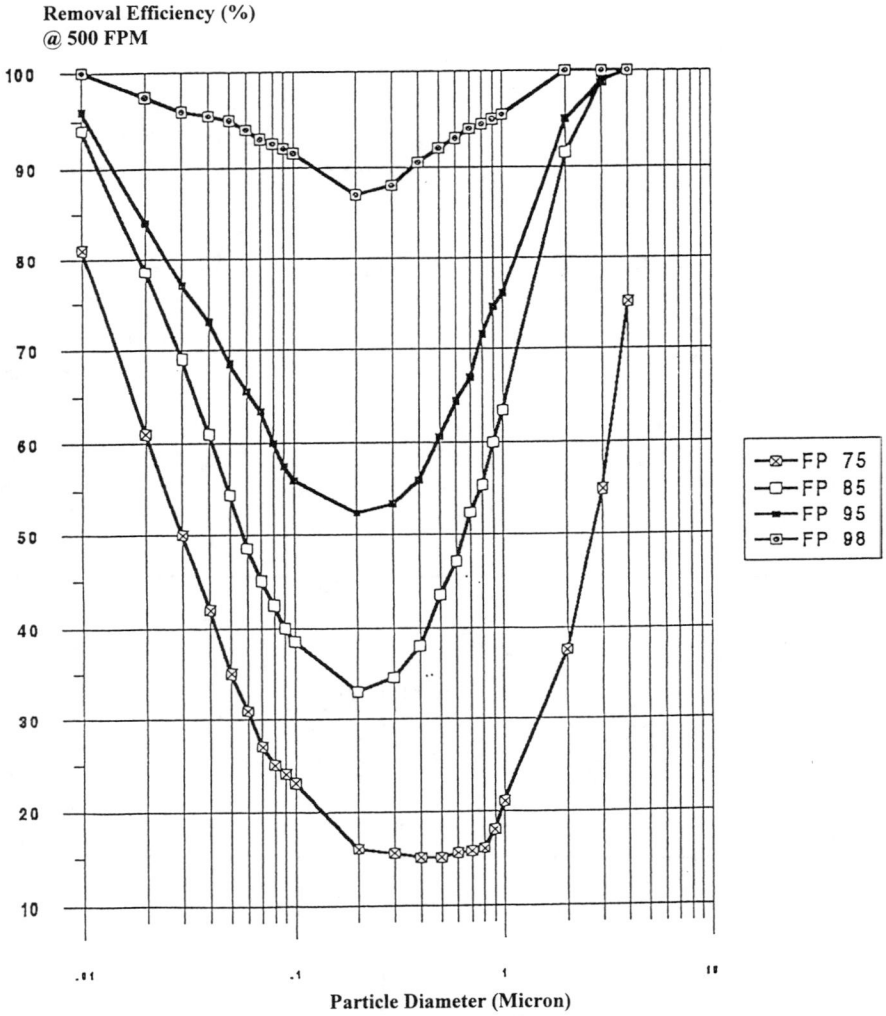

Figure 6-8. Most Penetrating Particle Sizes.

Obviously, conditions are less controlled since the challenge aerosol becomes the ambient air. However, the field performance of filtration banks can be evaluated in a similar fashion using the "real world" size fractions of the ambient airstream. Figure 6-9 shows the excellent fractional efficiency of a 95% ADST mini-pleat installation over time. The analysis uses the upstream and downstream particle counting for the determination. This can be compared to Figure 6-10 which typifies the field performance of a pocket filter also rated at 95% ADST.

6.7   GAS PHASE FILTRATION

The control technology for gaseous filtration is entirely different than particulate filtration. Particulate filtration is predominantly mechanical and/or electrical whereas gas-phase filtration is chemical. The primary capture device for chemical molecules is sorption. Sorption is a process based on the electron forces carried inherently with the molecule being attracted by similar forces on the surface of solid sorption filter beds. These "van der Waals" forces create powerful bonds by which molecules are attached to the surfaces of a solid matter (adsorption). Any solid surface is a dwelling place for "sorbed" molecules. However, a porous matter provides greater surface area and, thus, more sorption sites. The primary medium for gas-phase capture is a bed of extremely active or very highly porous sorption media such as an activated carbon. These materials have extremely high internal surfaces that can be measured as high as acres per ounce.

Gaseous control can also be accomplished with absorption. In fact, some sorbers such as zeolite and alumina are also very hydrophilic, attracting water at the same time as sorbing contaminant molecules. The combined process of adsorption while in contact with both water and other reactive molecules brings about ionization. This enables certain chemical reactions (chemisorption). To enhance this control mechanism, sorbents can be impregnated with specific reagents to enhance their ability to chemically react with gaseous contaminants of concern.

Another approach to attaining chemical reaction is the process of ionization. This employs an electromagnetic field which manipulates the electron charges of airborne molecules enabling them to react with each other. This process is usually employed in conjunction with a sorption bed, and these two processes have been demonstrated to act synergisti-

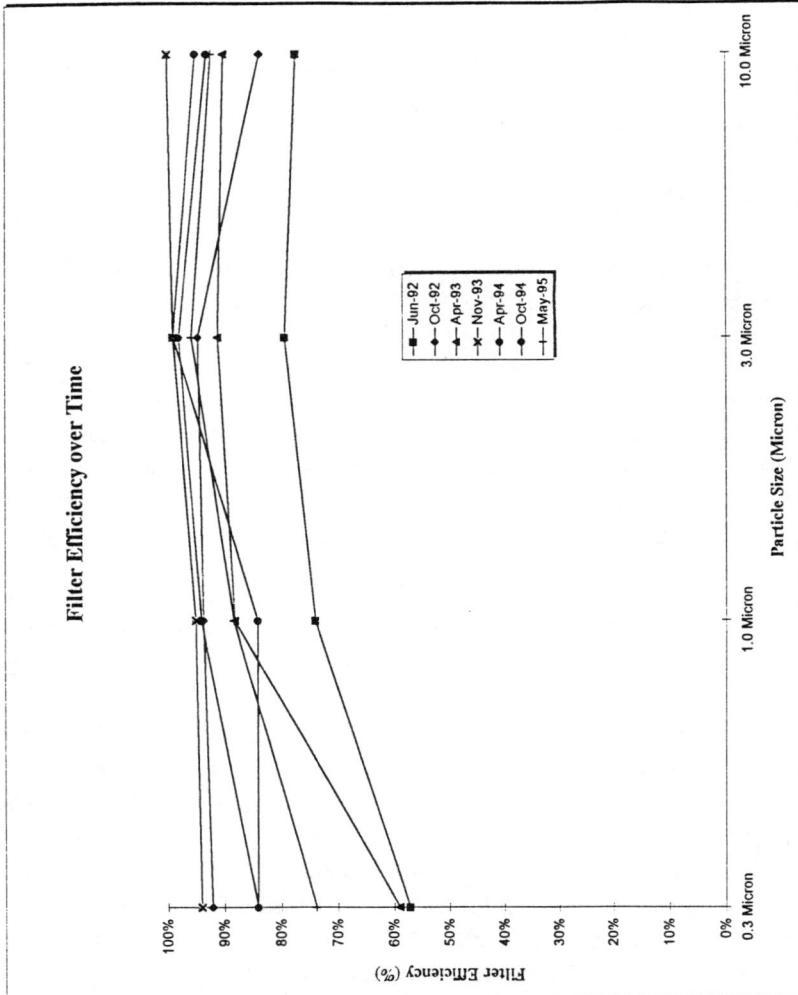

Figure 6-9. Fractional Efficiency of a 95% ADST Mini-Pleat Installation Over Time.

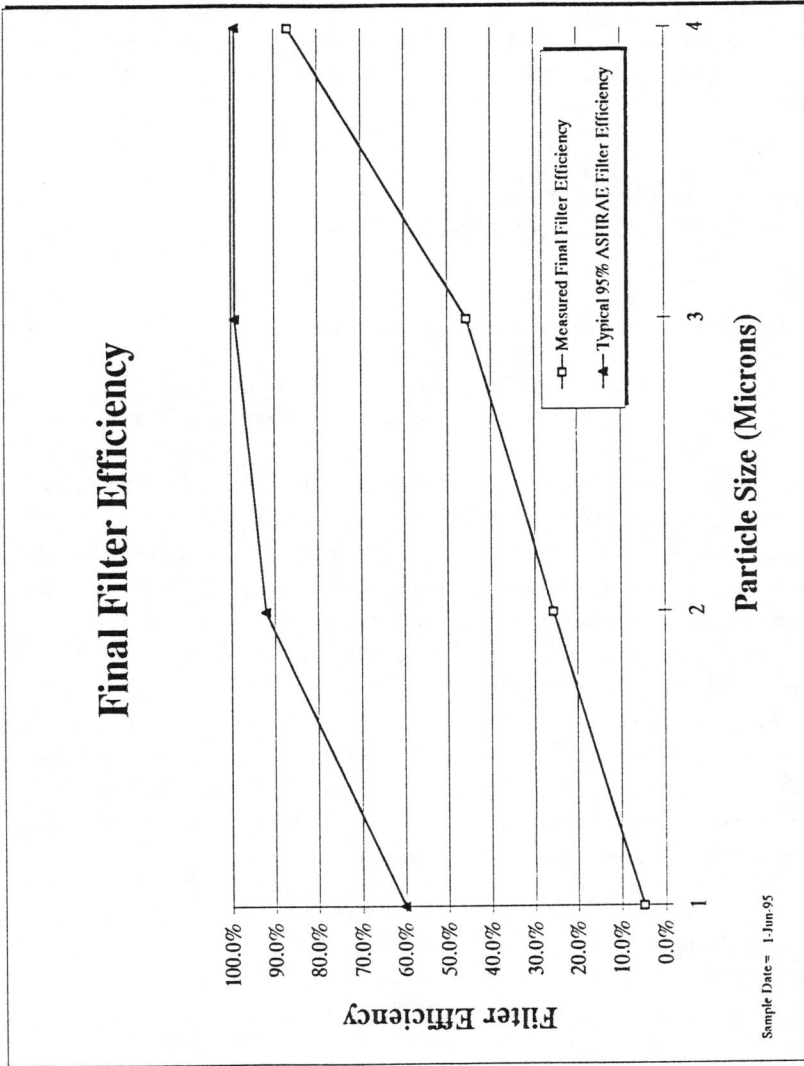

Figure 6-10. Filtration Efficiency of a Typical 95% ASHRAE Filter.

cally. Ozonization is also used for gas-phase control. Ozone is released into the airstream. It is highly unstable and, therefore, a highly reactive compound. Thus, it can chemically modify reactive molecules with which it comes in contact. This process must be used with a sorption bed for the synergy potential and to control the residual airborne ozone which is itself a regulated contaminant of health concern. The typical control media commercially available include:

- Carbon
- Impregnated carbon
- Alumina/potassium permanganate (KMnO$_4$)
- Zeolite (molecular sieve)
- Zeolite/KMnO$_4$
- Physical blends of the above-listed materials

The factors affecting efficacy of gaseous filters are much more complex than particulate filters, starting with the sorbent media properties. Generally, sorbents are in the form of either round cylindrical pellets or random-shaped flakes (refer to Figure 6-11). They can vary in internal surface area (generally the higher the surface area, the higher the capacity). Pellet size can affect apparent surface area as well as airflow and bypass characteristics. The inherent chemical properties of the sorbent can influence chemical contaminant preferences and control capabilities. Another physical property of concern is the flammability that can be critical when installed in an air conveyance system in a public building.

The contaminants themselves influence the nature and competence of the process. Unlike particulate matter that basically behaves the same (other than size), each gaseous contaminant is unique in its makeup and resulting behavior. The following are some of the factors that influence the behavior of specific chemical gasses in a sorption process:

- Contaminant
- Concentration
- Molecular weight
- Polarity
- Vapor pressure
- Acidity (pH)
- Reactivity
- Boiling point

Figure 6-12 typifies the variability of sorption performance as a function of varying challenge concentrations.

Similarly, atmospheric conditions like temperature, relative humidity (RH) and barometric pressure affect the performance of the sorption bed. Higher temperatures tend to make the molecules more active which tend to overpower the van der Waals forces. For example, carbon is normally reactivated with heat. High humidity can blind the pore structure, and void internal sorption sites by filling them with water molecules. These factors can have significant influence on the behavior of gaseous filters in the field. For example, a 10–°F shift or a 10-point RH swing can influence the sorption capacity isotherm of a carbon filter by as much as 25%.

The containment device for the sorption media will drastically affect the filter performance. This is where many leaps of logic occur in the application of gas-phase filtration. Because a sorbent is inherently capable of controlling a contaminant, the assumption is made incorrectly that control is true under all conditions of canister configuration (refer to Figure 6-13). Canister characteristics such as media bed depth, air velocity, and the resulting superficial dwell-time within the bed will drastically affect the performance of the filter. Generally, the higher the dwell-time,

Figure 6-11. Purafil Media with a Sample Bag and Probe.

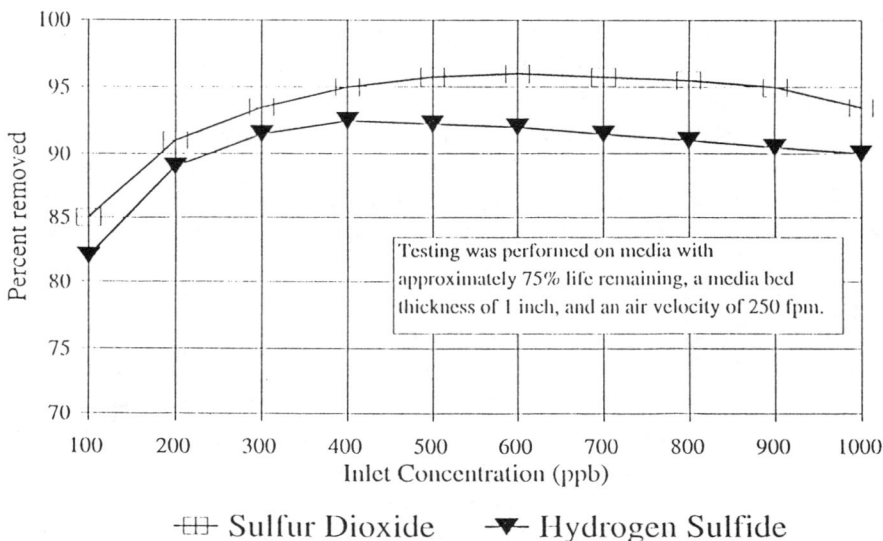

Figure 6-12. Single-Pass Efficiency of KMnO₄/Alumina Against Sulfur Dioxide and Hydrogen Sulfide.

the better opportunity for complete sorption.

Rating methods vary depending on the media but generally depend on proprietary data developed by the manufacturer. There is no current industry standard for rating the performance of the filter canister itself. Since each contaminant acts differently with each control media, it is difficult to have a universal test that applies to all conditions. ASHRAE is in the process of developing a standard, entitled: *ASHRAE Standard 145P*. Currently, the focus is on a laboratory method to standardize the evaluation and/or screening of the various media. A later draft of this standard will address full scale filter canister devices. Current rating methods include:

- Carbon Tet No.
- Isotherms
- Single point evaluations

- Single contaminant curve
- ASHRAE Standard 145 (pending)
- Physical properties

Since there are no universal test methods for rating and selecting sorbent product, the following anecdotal discussion may be helpful in the selection process.

Figure 6-13. Typical Air Cleaner Canisters.

Carbon is applied best against contaminants that are higher in concentration, heavier in molecular weight, nonpolar, and having a high-molecular-carbon content. This includes most volatile organic compounds (VOCs) and long-molecular-chain hydrocarbons. It also prefers lower atmospheric temperature and humidity. The $KMnO_4$-modified substrates perform best against lower-molecular-weight compounds, polar-compounds such as formaldehyde, and reactive inorganic compounds such as acid gases like hydrogen sulfide (H_2S) and sulfur dioxide (SO_2). Zeolite is particularly effective as a cation exchange media. This makes it perform well against contaminants like ammonia and other nitrogen-bearing compounds. It also has higher surface area than alumina compounds. Figure 6-14 shows a typical side access air-cleaner unit.

6.8   APPLICATION OF FILTRATION

The following decision paradigm will provide the designer with an effective process for applying filtration equipment.

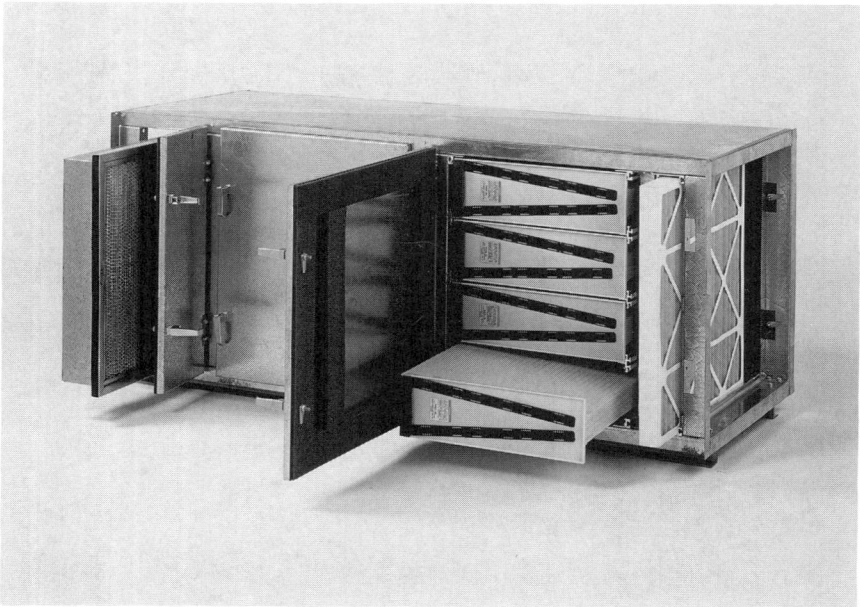

Figure 6-14. A Typical Side Access Air-Cleaner Unit.

6.8.1 Identify contaminant(s)

Usually the air is a complex gas that includes numerous contaminants. The nature and content of the complete airstream must be known to develop an appropriate system and select proper filtering media. This should include the contaminant nature, concentration, toxicity and health risk. Also, other atmospheric properties of the airstream must be understood to aid in the proper selection.

6.8.2 Establish control objectives

No perceptible odor, no more than 10,000 particles per cubic foot of respirable size, or no greater than 0.1 parts per million (ppm) H_2S for corrosion protection of electronic gear.

6.8.3 Select control strategy

Start with the selection of a filtering device or system. This includes the type, media and efficiency expectations.

6.8.4 Perform Life Cycle Cost Analysis (LCCA)

Before selecting a specific system or device, perform an LCCA. The initial selection may perform adequately, but LCCA may reveal short service life, resulting in higher maintenance cost for the owner.

6.8.5 Locate it in the system

Dependent on the contaminant source, type of mechanical system, site of exposure and control strategy.

6.8.6 Establish airflow path

This involves how the air will be delivered to and from the filtering device, how much air will be treated, and how it will be delivered to the targeted zone.

6.8.7 Build in space and capacity for retrofit

This allows for what may happen in the future. Perhaps the ultimate activity is unknown, the number of tenants are undetermined, or future expansions.

6.8.8 Evaluate use of prefiltration

According to some experts, most extended media particulate filters should not be prefiltered. This is contrary to the generally held belief that prefiltration protects the final filter, extending its useful life proportionately. Anecdotal experience indicates that the additional cost, room, labor, and energy consumption do not cost-justify the marginal gain of filter life. However, it is well advised to protect by prefiltration the following: HEPA, gas-phase sorbers, mini-pleat and EACs. The use of prefilters on the other filter types only increases cost, labor, energy, space, and adds little in extended life or value.

6.8.9 Install adequate filtration during construction

The filters of design level should be installed when the contamination and/or exposure levels occur at their highest peaks. Thus, the filtration system should be installed during the construction period when a major portion of the building distribution system occurs.

6.8.10 Provide for monitoring performance of filtration system

This can be done in situ with monitoring equipment that can signal failure or service needs to the building management control system. Periodic evaluation of the space will best ensure that a high level of air quality is sustained in the space.

6.9   ACKNOWLEDGMENTS

The author gratefully acknowledges the valuable contributions of the following entities: The Filtration Group, Inc., Joliet, IL, for their contribution of proprietary laboratory test data and generic product photographs; Environmental Design International, Ltd., Atlanta, GA, for proprietary field test data and critical contributions in the illustrations; Purafil, Inc., Atlanta, GA, for their contributions of generic gas-cleaning photographs; National Air Filtration Association (NAFA), Washington, DC, for permission to reproduce essential drawings and charts; and Research Triangle Institute, Research Triangle Park, NC, for their permission to reproduce charts.

6.10 REFERENCES

[1]ANSI/ASHRAE Standard 52.1-1992, *Gravimetric and Dust-Spot Procedures for Testing Air-Cleaning Devices Used in General Ventilation for Removing*

Particulate Matter, American Society of Heating, Refrigerating, and Air-Conditioning Engineers, Inc., Atlanta, GA.

[2]NAFA. *Guide to Air Filtration*. National Air Filtration Association, Washington, D.C.

[3]ASHRAE Standard 52.2P (Draft), *Method of Testing General Ventilation Air-cleaning Devices for Removal Efficiency by Particle Size*, American Society of Heating, Refrigerating, and Air-Conditioning engineers, Inc., Atlanta, GA.

[4]Burroughs, H.E. Barney "The Usage of Filtration as Fulfillment of Acceptable Indoor and Optimal Energy Management," *Proceedings of the Association of Energy Engineers, World Environmental Engineering Congress, 1992*.

7

Variable-Air-Volume/Bypass Filtration System to Control VOCs

Milton Meckler, P.E.
President, The Meckler Group
Encino, California

7.1   INTRODUCTION

In the modern indoor environment, volatile organic compounds (VOCs) are of major concern and there is considerable debate over what levels are acceptable for long-term exposure. This has direct effect on the required amount of indoor air. A total VOC concentration of 0.16 mg/m^3, has been suggested as the allowable concentration[1]. A multifactorial exposure range may exist between 0.2 mg/m^3 and 3.0 mg/m^3. Above the level of 3.0 mg/m^3, discomfort is expected, and above 25 mg/m^3, toxic effects may appear. To date, the VOC guideline suggested above is the most stringent among many others.

Properly implemented demand control ventilation (DCV) strategies can provide the opportunity to maintain acceptable indoor air quality (IAQ) in accordance with ASHRAE Standard 62-1989 while offering significant energy savings. This chapter will introduce a DCV system equipped with an IAQ sensor that directly measures the concentration of

VOCs in an occupied space and accordingly modulates supply airflow rates to help provide acceptable IAQ, comfort and cost-effectiveness, especially for variable-air-volume (VAV) systems. In addition, system filter and bypass filter selection criteria for both retrofit and new building applications will be outlined.

7.2 VARIABLE-AIR-VOLUME/BYPASS FILTRATION SYSTEM

Quality of comfort is a common problem in offices that have only one thermostat for several rooms. The lack of uniform response to outdoor temperatures and IAQ needs that vary independently is often a major drawback of conventional single-zone comfort systems. The Variable-Air-Volume/Bypass Filtration System (VAV/BPFS) shown in Figure 7-1 employs an electronic control system that provides cost-effective and improved VAV comfort while responding to varying IAQ requirements. It employs four separate zones, each with its own individual thermostat. The thermostat in each zone allows the system air damper to monitor and control carefully the temperature of each zone that is most comfortable to its occupants by changing the air-supply rate in response to "net-demands" for heating or cooling. Each zone damper can also communicate on a single-twisted pair of wires with the central controller that monitors each of the zone dampers and the temperatures. The central controller also automatically provides bypass control through direct air flow monitoring to allow constant fan-speed operation.

An IAQ sensor measures the concentration of VOCs to reset independently the supply-air temperature leaving the air-handling unit (AHU), increasing or decreasing the air flow through a filter/air-cleaner assembly in the bypass duct to maintain satisfactory VOC concentrations. The sensor does not attempt to make any distinction among indoor air contaminant types. Each AHU in the VAV/BPFS can have one or more IAQ sensors. The IAQ sensor can be located in either the return-air duct (duct air quality sensor) or the room (room air quality sensor). When it is located in the return-air duct, it senses the average contaminant concentration of all four zones combined. If there is a considerable variation in the concentration of indoor air contaminants in any one of the four zones, it may be more advantageous to locate the IAQ sensor in a "critical zone" rather than in the return-air duct.

Since VAV/BPFS is a constant-volume fan system at the AHU with

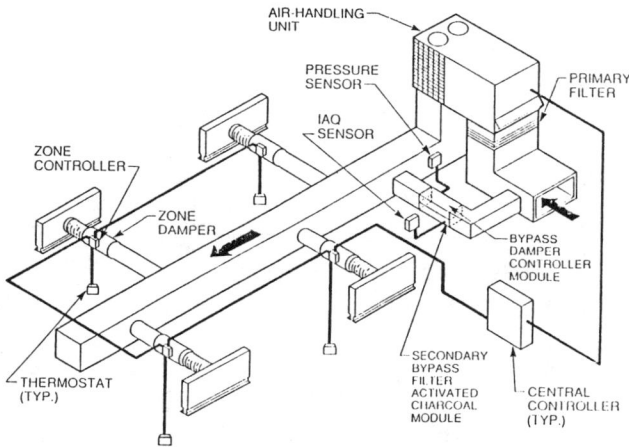

Figure 7-1. Variable-Air-Volume/Bypass Filtration System.

colder air delivered to all four zones, each zone thermostat will modulate its zone damper to close to match the increased cooling capacity of the supply air with the coincident space thermal load. If the AHU is in a heating cycle, the supply-air temperature will increase. If the AHU is in a cooling cycle, the supply-air temperature will decrease to reach the maximum permissible VOC concentration. When the zone dampers modulate to close further, the bypass sensor senses the pressure buildup and automatically allows more supply air to be bypassed through the VAV/BPFS system bypass duct to the high-efficiency filter/air-cleaner assembly.

For the purpose of analysis, Figure 7-2 models a constant flow heating, ventilating and air-conditioning (HVAC) system with a VAV bypass loop. To calculate the contaminant concentration in an occupied space where contaminants are generated, removed or added by air flows, one must perform a mass-balance of space volumes (V and V')[2]. The rate of change in total mass contaminant in the occupied zone V is expressed by the following:

$$V(dc/dt) = \dot{N}_{generated} + \dot{N}_{added} - \dot{N}_{removed} \tag{7-1}$$

Placing a high-efficiency filter in the VAV bypass loop causes a portion of the system constant air flow to be cleaned further, thus providing cleaner air to the occupied space.

To consider the effect of VAV bypass on the quality of supply air

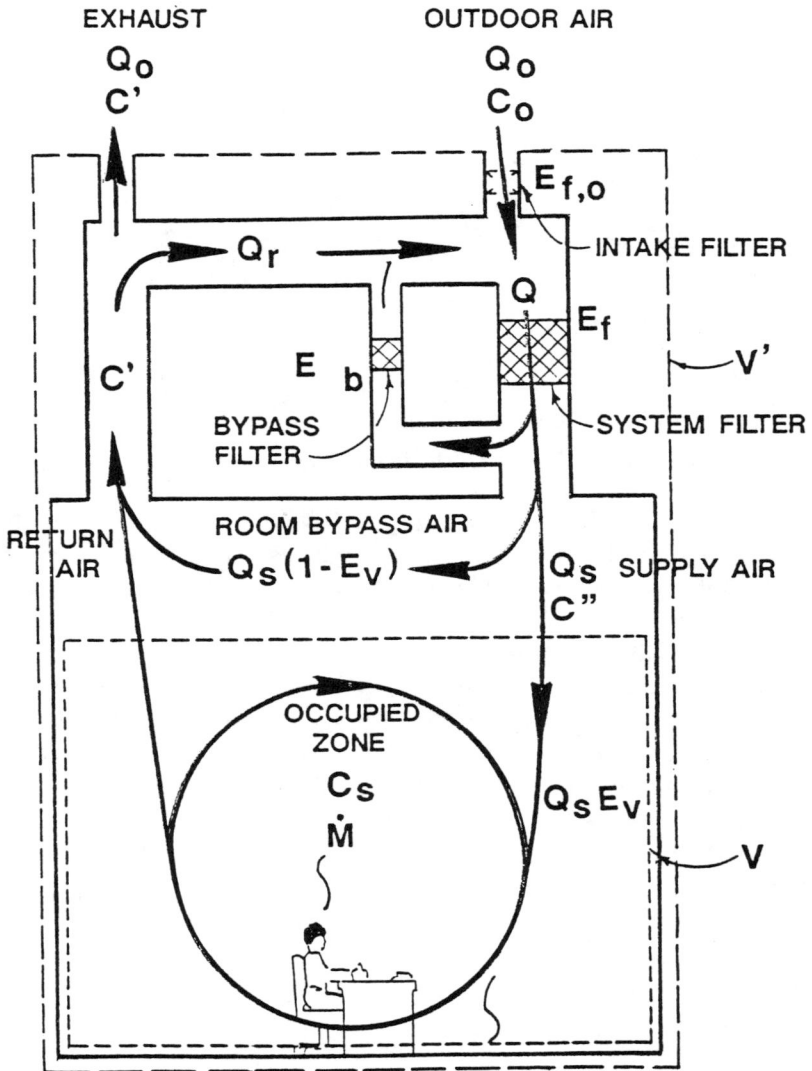

Figure 7-2. Constant-Volume HVAC System with a VAV-Bypass Loop.

(not the occupied zone air), one must compare the contaminant concentration of supply air for the system with (C_S) and without ($C''_{s,Qb} = 0$) a bypass loop It can be shown that Eqn. (7-2) below can be derived from Eqn. (7-1); for brevity, this derivation is omitted here. In Eqn. (7-2), the contribution of outdoor air contaminants to indoor air contaminants is assumed to be negligible.

$$C''_s/C''_{s,Qb}= 0 = [(Q_o/Q) + (1 - Q_o/Q)\,E_f]/$$
$$\{(1 - (Q_b/Q)[(Q_b/Q + (Q_b/Q)\,E_{f,b}$$
$$(1 - E_f) + (1 - (Q_o/Q)]E_f\} \qquad\qquad (7\text{-}2)$$

where

Q_o: outdoor airflow rate,

Q: system airflow rate,

Q_b: bypass airflow rate,

$E_{f,b}$:   bypass filter efficiency, and

E_f: system filter efficiency.

Eqn. (7-2) which is independent of ventilation effectiveness (E_v), is plotted in Figure 7-3 for $E_f = 0\%$[3,4]. Note that for a system having a low-efficiency system filter (e.g., without a gaseous-phase filter), there is also an optimum bypass airflow rate (e.g., 30% to 50% of constant Q) that results in cleaner supply air.

The supplemental air-cleaning strategy used here is attractive because of the energy savings achieved by a lower system filter pressure-drop and the lower capital cost associated with the air flow in the main HVAC system design. For an HVAC system with a VAV-bypass loop, whether a retrofit or a new building, the main system filter efficiency can be determined at a maximum system airflow rate instead of a minimum system airflow rate as in a VAV system without a bypass loop[2]. Selected particulate filters must always be used in conjunction with gaseous-phase carbon adsorbers to remove VOCs[5]. Particulate filters are rated in accordance with their efficiencies on a mass-mean-diameter (MMD) of 0.3 micron, and carbon adsorbers are based on the VOC toluene.

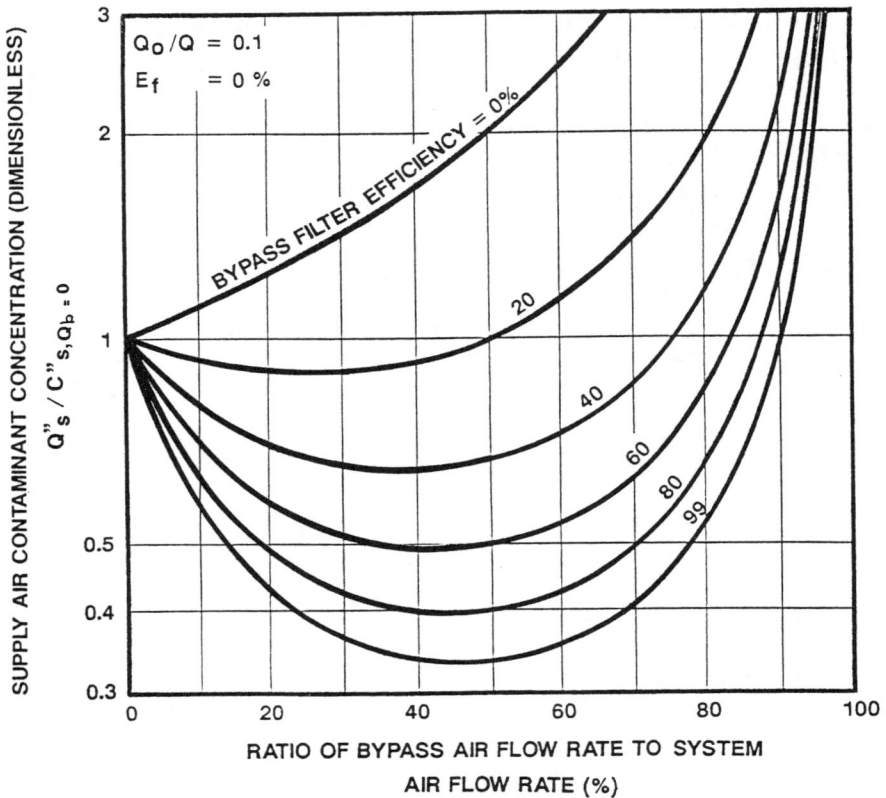

Figure 7-3. Effect of VAV-Bypass Loop on Supply Air Quality (Independent of Ventilation Effectiveness).

7.3 FILTER SELECTION CRITERIA
FOR RETROFIT AND NEW BUILDINGS

To illustrate the use of this filter selection criterion, consider the following example applied to a retrofit and a new building. Suppose the designer wishes to satisfy the minimum outdoor airflow rate of 20 cubic feet per minute (cfm)/person for a retrofit office building (recommended by ASHRAE Standard 62-1989) using a reduced outdoor airflow rate of 5 cfm/person.

Further assume that the building has a VAV system without a by-pass loop, a supply airflow rate (V_s) of 50 cfm/person, and $E_v = 65\%$, a typical office building. Using Figure 7-4[3,4,6], the designer locates the particulate and adsorption efficiency of 45% for V_s = 50 cfm/person (0.35 cfm/ft^2). The system supply airflow rate of 0.35 cfm/ft^2 is based on the recommended 7 people/1000 ft^2 for an office occupancy per ASHRAE Standard 62-1989. Then, using Figure 7-5 for an MMD of 0.3 micron particulate size and the particulate and adsorption efficiency of 45%, the designer selects Filter C as the system filter.

Now, suppose the designer again wishes to satisfy the minimum outdoor airflow rate of 20 cfm/person using a reduced outdoor airflow rate of 5 cfm/person for the same retrofit building having a VAV system

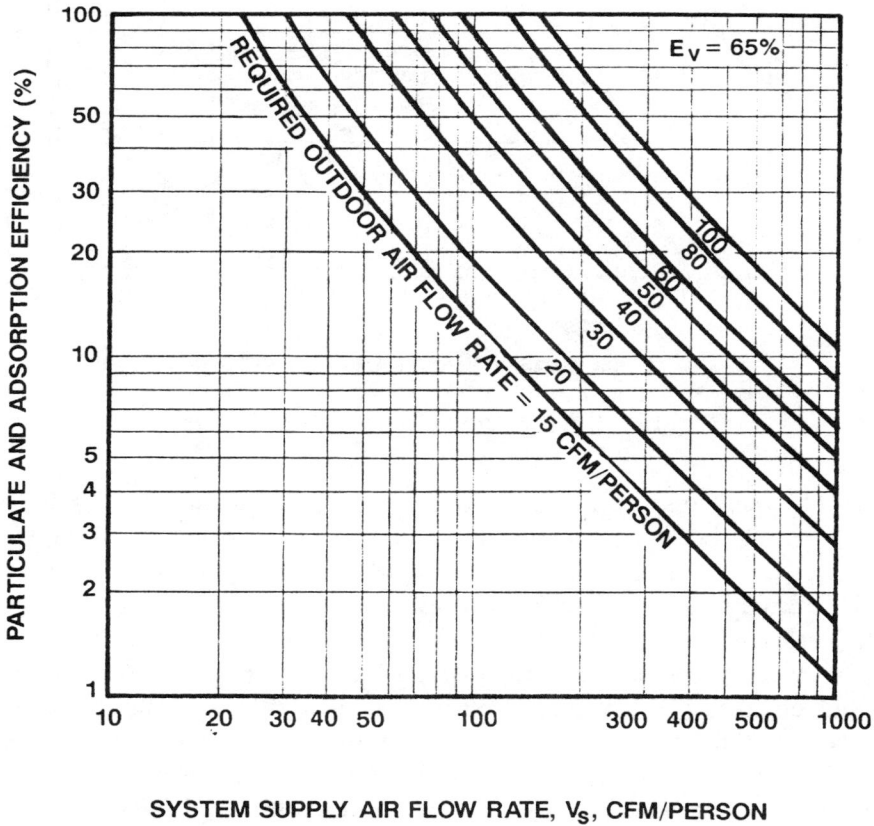

SYSTEM SUPPLY AIR FLOW RATE, V_s, CFM/PERSON

Figure 7-4. Determination of System Filter Efficiency with Reduced Outdoor Airflow Rate of 5 cfm/person for Retrofit Buildings.

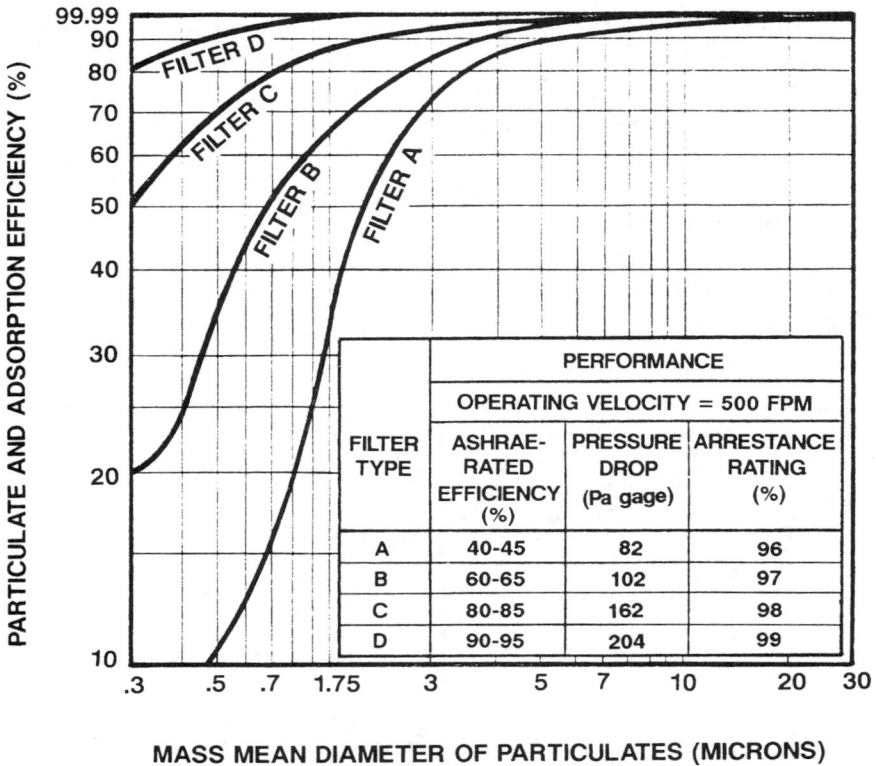

The graph shows PARTICULATE AND ADSORPTION EFFICIENCY (%) on the vertical axis (from 10 to 99.99) versus MASS MEAN DIAMETER OF PARTICULATES (MICRONS) on the horizontal axis (from .3 to 30), with curves for FILTER A, FILTER B, FILTER C, and FILTER D.

FILTER TYPE	PERFORMANCE		
	OPERATING VELOCITY = 500 FPM		
	ASHRAE-RATED EFFICIENCY (%)	PRESSURE DROP (Pa gage)	ARRESTANCE RATING (%)
A	40-45	82	96
B	60-65	102	97
C	80-85	162	98
D	90-95	204	99

MASS MEAN DIAMETER OF PARTICULATES (MICRONS)

Figure 7-5. A Typical Filter Selection Criteria.

with a bypass loop. This time, assume V_s = 90 cfm/person and again E_v = 65%. Similarly, using Figure 7-4 for V_s = 90 cfm/person the designer locates the corresponding particulate and adsorption efficiency of 20%. Accordingly, using Figure 7-5 once again, for the particulate and adsorption efficiency of 20%, the designer selects Filter B as the system filter.

Most recent mathematical analyses have shown that if the required filter efficiency for the main system is relatively low, placing a high-efficiency filter in the bypass loop can improve IAQ, provided the bypass airflow rate does not exceed approximately 30% of the main system airflow rate[7]. This is graphically illustrated in Figure 7-6[3,4] for a reduced outdoor airflow rate for retrofit buildings. One can see that, if the bypass filter efficiency is less than 80%, the contaminant concentration in the occupied zone (shown as a dimensionless ratio in Figure 7-6) will be

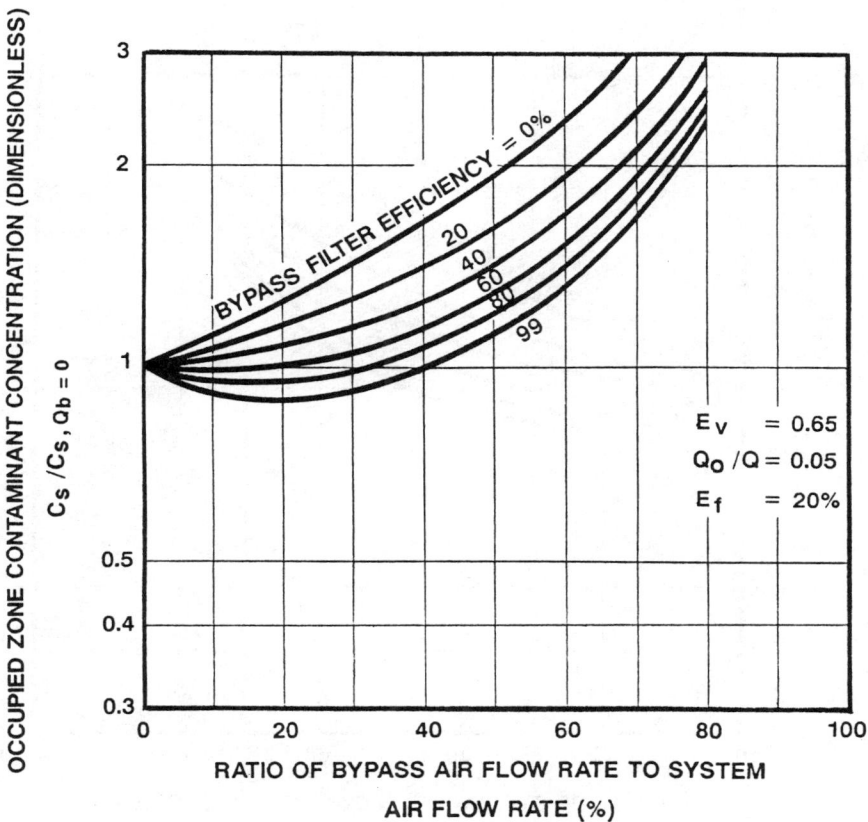

Figure 7-6. Effect of VAV-Bypass Filter for Retrofit Buildings.

greater than unity with increased bypass airflow rates and, therefore IAQ will suffer. On the other hand, if the bypass filter efficiency is greater than 80%, the contaminant concentration in the occupied zone will be approaching unity. As a result of this, IAQ will improve with reduced supply airflow rates, provided the bypass airflow rate does not exceed approximately 30% of the main system airflow rate. Similarly, the effect of placing a high-efficiency filter in a bypass loop is also illustrated in Figure 7-7 for new buildings. For the retrofit building above, the designer selects Filter D (using Figure 7-5) as the bypass filter with a corresponding bypass filter efficiency of 80% as shown in Figure 7-6.

Let us now consider the following example applied to a new building. Suppose the designer again wishes to satisfy the minimum outdoor

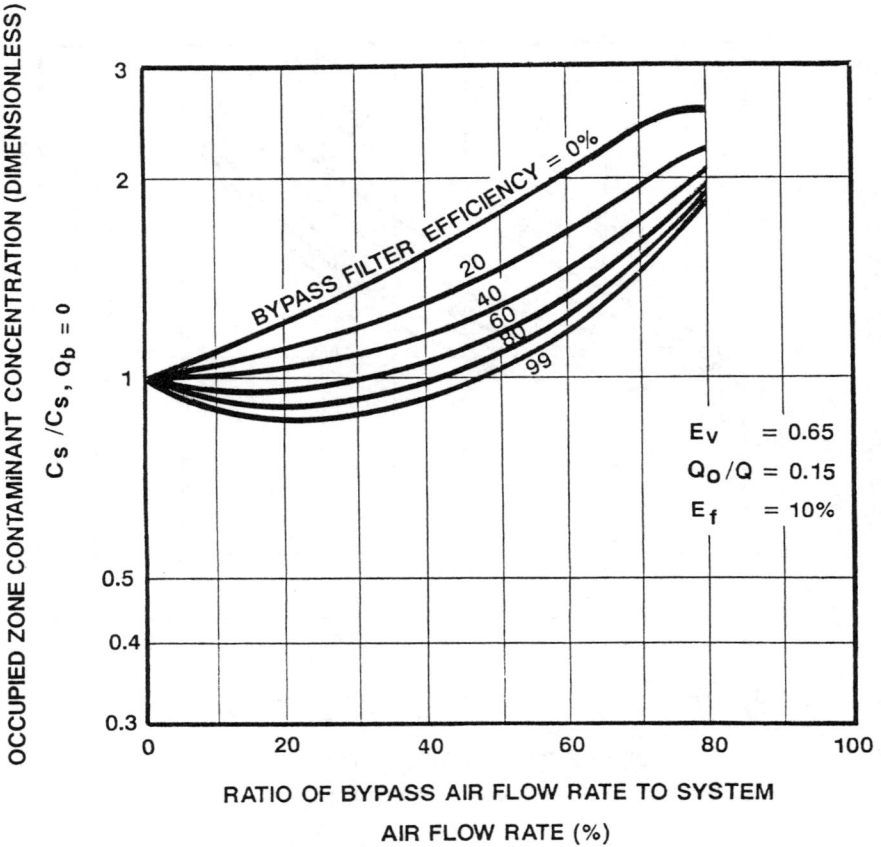

Figure 7-7. Effect of VAV-Bypass Filter for New Buildings.

airflow rate of 20 cfm/person using a reduced outdoor airflow rate of 15 cfm/person. Further assume that the new building has a VAV system without a bypass loop, V_s = 50 cfm/person and E_v = 65%. Using Figure 7-8, the designer locates the particulate and adsorption efficiency of 30% and selects Filter C as the system filter (refer to Figure 7-5).

For the case of a new building having a VAV system with a bypass loop and V_s = 90 cfm/person, the designer locates the particulate and adsorption efficiency of 10% in Figure 7-8[3,4,6]. Using Figure 7-5, the designer selects Filter A as the system filter. Similarly, using the same selection criteria as for the retrofit building having a VAV system with a bypass

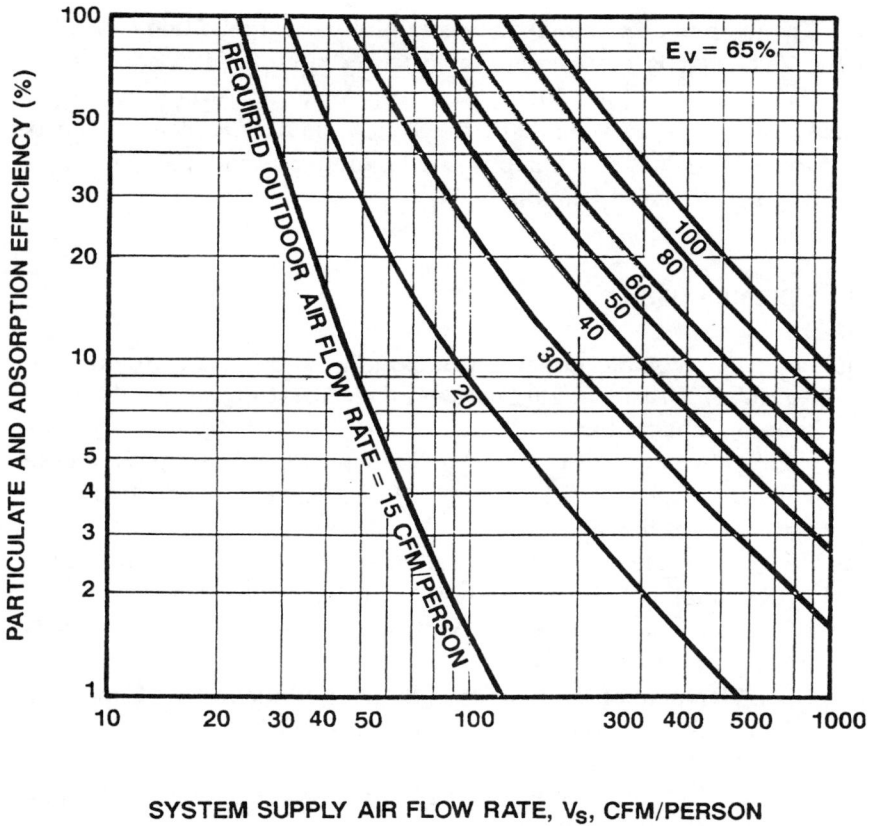

The chart axes:
- Y-axis: PARTICULATE AND ADSORPTION EFFICIENCY (%), ranging from 1 to 100
- X-axis: SYSTEM SUPPLY AIR FLOW RATE, V_S, CFM/PERSON, ranging from 10 to 1000
- $E_V = 65\%$
- Diagonal label: REQUIRED OUTDOOR AIR FLOW RATE = 15 CFM/PERSON
- Curve labels: 20, 30, 40, 50, 60, 80, 100

Figure 7-8. Determination of System Filter Efficiency with Reduced Outdoor Airflow Rate of 15 cfm/person for New Buildings.

loop, the designer selects again Filter D as the bypass filter with a corresponding bypass filter efficiency of 80% as shown in Figure 7-7.

7.4   CONCLUSIONS

The reduction of supply airflow rate can adversely affect IAQ in an occupied space of buildings with VAV systems. Use of a supplemental high-efficiency filter/air-cleaner assembly in the VAV-bypass loop can provide a means for offsetting reduced air supply rates without sacrificing

IAQ. The temperature of each zone is carefully monitored and controlled to the level that is most comfortable to the occupants. Cleaner air is supplied to the occupied zone whenever the supply airflow rate is reduced during the VAV-mode of the system, and VOC control can be achieved within 30% of the supply air flowrate and independent of the space temperature and humidity.

7.5   REFERENCES

[1]Molhave, L., "Volatile Organic Compounds, Indoor Air Quality and Health," *Fifth International Conference on Indoor Air Quality and Climate,* Toronto, Canada, 1990.

[2]Yu, H.H.S., "Analysis of Indoor Air Quality with VAV-Bypass Supplemental Air Cleaning," *Proceedings of First Annual Indoor Air Quality Exposition,* Tampa, FL., 1992.

[3]Yu, H.H.S., "Supplemental Air Filtration and Indoor Air Quality," Pt. 1, FARR Company Laboratory Report 1409-2 Add. 13, 1991.

[4]Liu, R.T., et al., "Filtration and Indoor Quality," FARR Company Laboratory Report 1409-6, Rev. B, 1991.

[5]Liu, R.T., "Removal of Volatile Organic Compounds in IAQ Concentrations with Short Carbon Bed Depths," *Fifth International Conference on Indoor Air Quality and Climate,* Toronto, Canada, 1990.

[6]Liu, R.T., et al., "Filter Selection on an Air Engineering Basis," *Heating/Piping/Air Conditioning,* 1991.

[7]Yu, H.H.S., and R.R. Raber, "Implications of ASHRAE Standard 62-1989 on Filtration Strategies and Indoor Air Quality and Energy Conservation," *Proceedings of Fifth International Conference on Indoor Air Quality and Climate,* Toronto, Canada, 1990.

8

Testing of Variable-Air-Volume/ Bypass Filtration System for Improved IAQ

Demetrios J. Moschandreas, Ph.D.
Professor
Illinois Institute of Technology
Chicago, Illinois
and
S.W. Choi, Ph.D.
Illinois Institute of Technology
Chicago, Illinois

8.1   INTRODUCTION

Conventionally designed constant-volume (CV) and variable-air-volume (VAV) systems both have their inadequacies. The shortcomings of a CV system include inefficient energy consumption[1,2] and the disadvantages of a VAV system include increased indoor air contaminant concentrations because of inadequate ventilation[3,4]. Therefore, a heating, ventilating and air-conditioning (HVAC) designer today is faced with an important and a difficult decision in selecting an HVAC system that must provide cost-effective energy consumption and acceptable indoor air quality (IAQ) in accordance with ASHRAE Standard 62-1989.

The Variable-Air-Volume/Bypass Filtration System (VAV/BPFS) is designed to rectify inadequacies of both the CV and VAV systems. The VAV/BPFS system is a variation of the conventional VAV system with a demand controlled high-efficiency filtration/air-cleaning system. It responds to thermal loads and/or indoor air contaminant concentration levels in a conditioned space providing cost-effective thermal comfort and acceptable IAQ. In this chapter, we will present the results of a study conducted at Illinois Institute of Technology, Chicago, Illinois under controlled conditions to investigate whether the VAV/BPFS system (introduced in Chapter 7 of this book) improves IAQ by decreasing the concentrations of indoor air contaminants, and reduces energy consumption in comparison to a conventionally designed VAV system.

8.2   TEST PROCEDURE

All experiments were performed in an 1150-ft^3 room size aluminum chamber in cooling mode. This structure was designed to control temperature, relative humidity, air recirculation and air-exchange rate. During the performance of experiments, two indoor air contaminant sources (commercially available cigarettes and spray deodorant) and energy loads of 400, 800, 1200 and 1900 watts were used. Energy loads simulated the presence of occupants in the chamber (a space to be conditioned), and induce the operation of the ventilation system. An energy load of 200 watts simulates the presence of one occupant in an occupied space[5].

Twelve experiments were performed under controlled chamber conditions lasting 100 minutes. All experiments were performed in the cooling mode. The chamber setpoint temperature was assigned the value of 75°F. The test chamber was properly modified to help measure both the IAQ parameters and energy parameters during the operation of either the conventionally designed VAV system or the VAV/BPFS system. Modification made to the original VAV system allowing to select the HVAC mode of operation between VAV and VAV/BPFS systems were very easy and relatively inexpensive. Figure 8-1 shows the VAV/BPFS system. The experiments measured the infiltration rate in the chamber for the first 40 minutes; and measured outdoor supply airflow rate, total effective removal rates (v_{TEM}) of carbon monoxide (CO), total volatile organic compounds (TVOCs) and particulate matter (PM); and energy consumption for the next 50 minutes at each energy load.

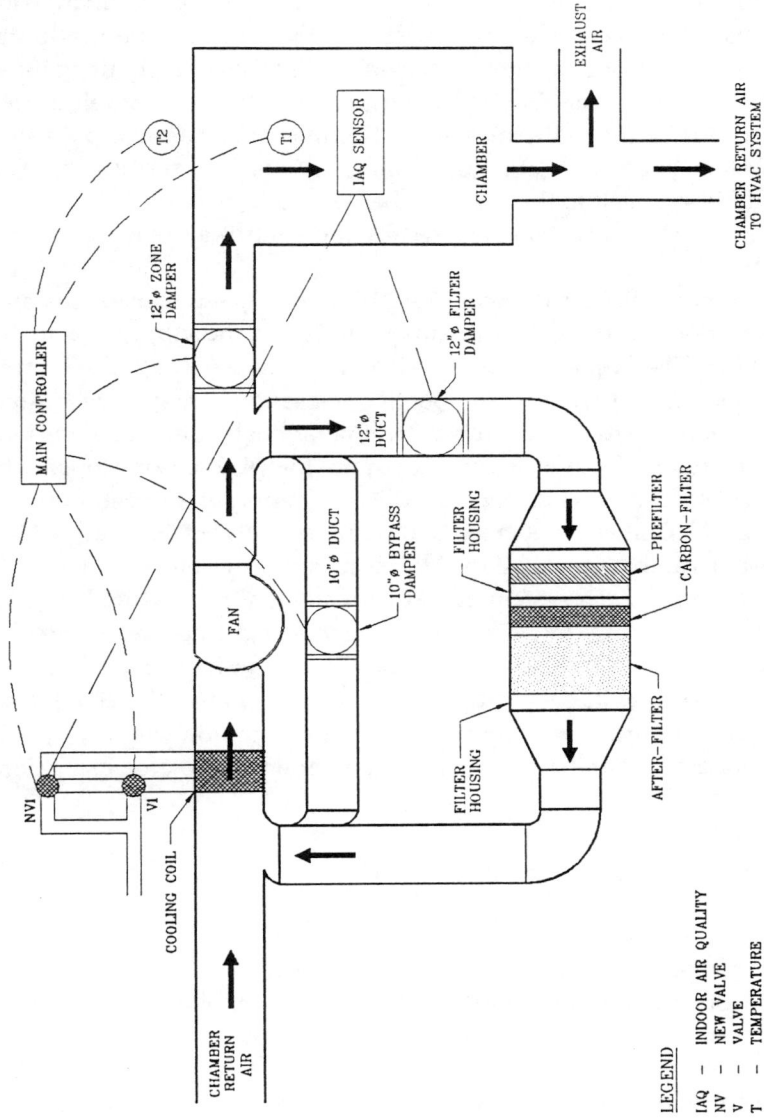

Figure 8-1. Variable-Air-Volume/Bypass Filtration System.

LEGEND

IAQ – INDOOR AIR QUALITY
NV – NEW VALVE
V – VALVE
T – TEMPERATURE

The VAV system (shown in Figure 8-1) is used in conjunction with a filtration system to increase energy savings by reducing the outdoor airflow rate, and to improve IAQ by reducing indoor air contaminant concentrations. The control system utilizes two thermostats, one inside the chamber (T1) and one outside the chamber (T2). A 12-inch air volume damper (AVD) (or zone damper) controls the air flow by an electronic controller responding to thermostat T1. Thermostat T2 has an input value to the system to create heating or cooling demands on the system. The cooling air is supplied to the test chamber via a cold-water coil controlled by Valve V1. This valve is energized by the control system mentioned above.

A 10-inch relief air damper (RAD) (or bypass damper) opens and closes in response to the static-pressure variations in the supply duct to the test chamber. The 12-inch AVD filter damper of the VAV/BPFS system is controlled by the IAQ sensor, and opens or closes on demand as required by the air quality in the test chamber. The operational difference of the two systems, therefore, focuses on whether or not the IAQ sensor is operated. The IAQ sensor controls Valve V1 to reduce the supply air temperature and the 12-inch AVD filter damper, so that more recirculated air of the VAV/BPFS system can pass through the bypass filtration system.

The effect of each system on IAQ quantities was measured by estimating the v_{TER} of CO, TVOC and PM. The v_{TER} is defined as the ventilation rate required to reduce the concentration of an indoor air contaminant to the level that its concentration is reduced by the combined effect of infiltration rate, outdoor supply airflow rate and filtration rate. The v_{TER} of each indoor air contaminant in the test chamber was calculated using Eqn. (8-1):

$$C_{t=t} = C_{t=0} \exp(-v_{TER} t) \tag{8-1}$$

where

$C_{t=0}$:   contaminant concentration at t=0 (mass/volume),

$C_{t=t}$:   contaminant concentration at t=t (mass/volume), and

t: time (h^{-1}).

The concentrations of CO, TVOC and PM were measured in the chamber center and outside of the chamber every 10 minutes.

Energy quantities, temperatures and airflow rates were measured to estimate and compare energy consumption of the two systems. Temperatures were measured at the following chamber HVAC system sites: makeup air, return air, before chilled-water system, after chilled-water system, mixed air, supply chilled-water and return chilled-water. The make-up airflow rate, supply airflow rate, bypass damper airflow rate, filter damper airflow rate and recirculated airflow rate were measured at appropriate sites of the chamber and HVAC system. The total energy consumption (TEC) was calculated by the Eqn. (8-2) below:

$$TEC = E_{fan} + E_{pump} + E_{chiller} \qquad (8\text{-}2)$$

where
E_{fan}: energy of fan,
E_{pump}: energy of chiller-water pump, and
$E_{chiller}$: energy of chiller.

The concentrations of CO, TVOC, PM and temperatures and airflow rates were measured by the instruments. A previous study[6] showed that at the height of measurement the contaminant concentrations are uniformly distributed throughout the chamber. Calibration of the instrument measuring the concentrations of PM was checked with the zero control and a secondary calibration once each experimental day. Daily calibrations of the instrument measuring the concentrations of TVOC were performed using a tracer-gas analyzer. Quality control of the instruments for airflow rates and temperature measurements were performed on a regular basis.

8.3   STUDY RESULTS

Figure 8-2 shows the estimated outdoor supply airflow rates for the conventional VAV and VAV/BPFS systems as a function of energy loads used. For each energy load, the difference between outside supply airflow rates (corresponding values from the VAV and VAV/BPFS systems) were compared. The "null hypothesis" (Ho : $v_{OSA.VAV} = v_{OSA.VAV/BPFS}$) was rejected at significance level, $\alpha = 0.05$. The difference between the outside supply airflow rates of VAV and VAV/BPFS systems was statistically significant, and the $v_{OSA.VAV/BPFS}$ was lower than $v_{OSA.VAV}$. The chamber infiltration rate was measured during the first 40 minutes of each experiment. During this 40-minute period of time, there was no mechanical ventilation and the removal of CO was due to dilution by air infiltrated into the test chamber. As expected, the chamber infiltration rate (v_i) was nearly constant in all experiments. The v_i ranged from 0.28 ach (air changes per hour) to 0.31 ach.

The null hypothesis ($H_0 : v_{i.VAV} = v_{i.VAV/BPFS}$) of no difference among the infiltration rates could not be rejected ($\alpha = 0.05$), across the energy loads of 400 watts to 1900 watts. The total effective removal rate of CO ($v_{TER.CO}$) is used to estimate the mechanical ventilation rate (v_m). The $v_{TER.CO}$ is composed of v_i and v_m. Therefore, the v_m is calculated by Eqn. (8-3) below:

$$v_m = v_{TER.CO} - V_i \qquad\qquad (8\text{-}3)$$

The v_m increased from 1.55 ach to 3.43 ach with increasing energy loads (400 watts to 1900 watts). This resulted in the introduction of more outside air into the test chamber.

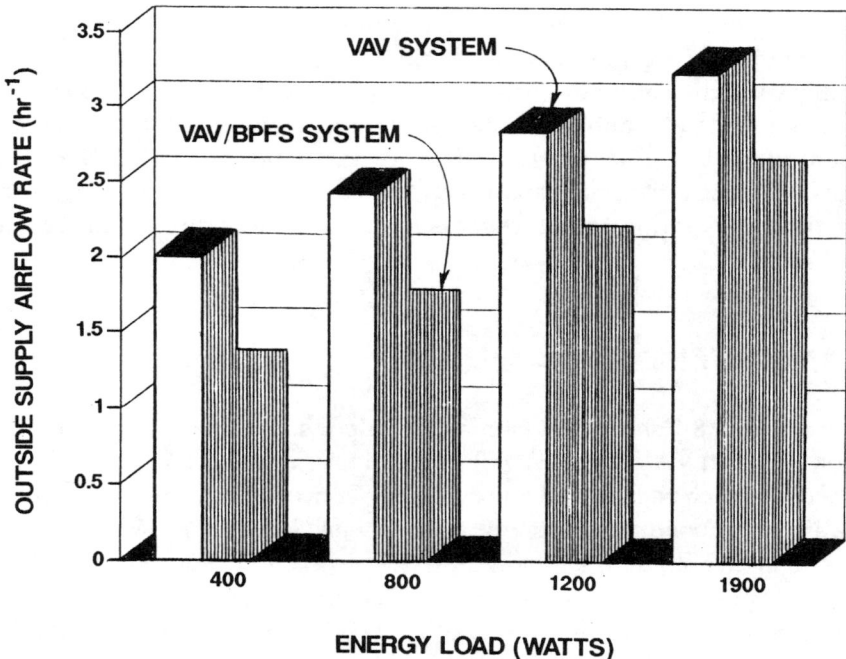

Figure 8-2. Outside Supply Airflow Rates vs. Energy Loads for VAV and VAV/BPFS Systems.

The total effective removal rate of PM ($v_{TER.PM}$) was increased as the energy loads increased (refer to Table 8-1). A statistical analysis conducted for each energy load revealed that the null hypothesis of no difference in removing PM by each system (Ho : $v_{TER.PM.VAV}$ = $v_{TER.PM.VAV/BPFS}$) is rejected at significance level, α = 0.05, and that the PM removal rate increased using the VAV/BPFS system (HA: $v_{TER.PM.VAV}$ < $v_{TER.PM.VAV/BPFS}$). Similarly, the total effective removal rate of TVOC ($v_{TER.TVOC}$) was statistically more significant with the VAV/BPFS system (refer to Table 8-2). These findings indicate that contaminant removal with the VAV/BPFS system is more efficient than the VAV system.

The $v_{TER.PM}$ and $v_{TER.TVOC}$ both consist of chamber infiltration rate (v_F), v_m and v_i. The VAV/BPFS system filtration effect on PM ($v_{F.PM.VAV/BPFS}$) and TVOC ($v_{F.TVOC.VAV/BPFS}$) are calculated using the Eqn. (8-4) and Eqn. (8-5) below:

$$v_{F.PM.VAV/BPFS} = v_{TER.PM} - v_i - v_m \qquad (8\text{-}4)$$

$$v_{F.TVOC.VAV/BPFS} = v_{TER.TVOC} - v_i - v_m \qquad (8\text{-}5)$$

As shown in Table 8-3, $v_{F.PM.VAV/BPFS}$ and $v_{F.TVOC.VAV/BPFS}$ increased from 3.70 ach to 5.97 ach and 0.73 ach to 1.20 ach, respectively, as the energy load increased from 400 watts to 1900 watts. The time variations of average CO, PM and TVOC concentrations with each energy load are shown in Figure 8-3.

Bernoulli's Theorem states that the impact pressure at any point in a duct is equal to the impact pressure at any other point. Thus, total pressure (velocity pressure + static pressure) is constant throughout the length of the duct if energy is neither gained nor lost. Figure 8-4 shows the application of the Bernoulli's Theorem to both the VAV and VAV/BPFS systems at pressure measurements points. Observed values of total pressure at the measurement points satisfied Bernoulli's Theorem within a 5% experimental error. A quantitative pressure distribution curves throughout the supply-air duct for the VAV and VAV/BPFS systems are shown in Figure 8-5. The relative pressure-drop between the outside inlet and the left side of the fan is more negative for the VAV system than the VAV/BPFS system. This should help explain the observed outside supply airflow rates that are reduced.

The positions of individual zone dampers constantly change in both systems, depending on the chamber cooling demand. Since the supply-air

Table 8-1. Total Effective Removal Rate of PM.

ENERGY LOAD (Watts)	NUMBER OF EXPERIMENTS	TOTAL EFFECTIVE REMOVAL RATE (hr^{-1})	
		$V_{TER,PM,VAV}$	$V_{TER,PM,BPFS}$
400	3	2.10 ± 0.10 *	5.53 ± 0.06
800	3	2.87 ± 0.15	6.80 ± 0.26
1200	3	3.63 ± 0.25	7.97 ± 0.23
1900	3	4.53 ± 0.32	9.70 ± 0.17

* MEAN $\pm$ S.D.

Table 8-2. Total Effective Removal Rate of TVOC.

ENERGY LOAD (Watts)	NUMBER OF EXPERIMENTS	TOTAL EFFECTIVE REMOVAL RATE (hr^{-1})	
		$V_{TER,TVOC,VAV}$	$V_{TER,TVOC,BPFS}$
400	3	2.57 ± 0.16 *	3.03 ± 0.06
800	3	2.80 ± 0.10	3.60 ± 0.17
1200	3	3.20 ± 0.10	4.00 ± 0.10
1900	3	3.60 ± 0.10	4.47 ± 0.06

* MEAN $\pm$ S.D.

Table 8-3. VAV/BPFS Filtration Effect on PM and TVOC.

ENERGY LOAD (Watts)	NUMBER OF EXPERIMENTS	VAV/BPFS FILTRATION EFFECT	
		$V_{F.PM. VAV/BPFS}$	$V_{F.TVOC. VAV/BPFS}$
400	3	3.70 ± 0.00	0.73 ± 0.06
800	3	4.27 ± 0.32	0.70 ± 0.10
1200	3	4.67 ± 0.25	1.07 ± 0.12
1900	3	5.97 ± 0.15	1.20 ± 0.00

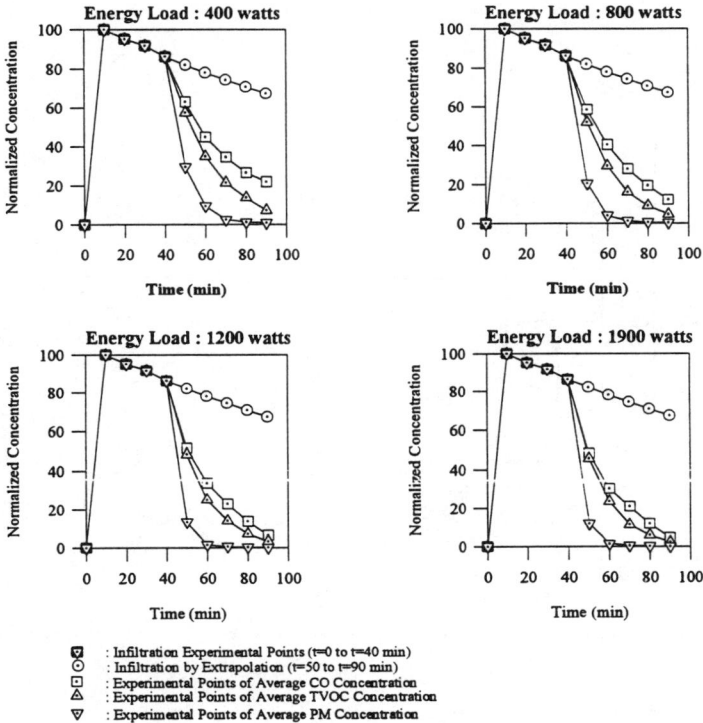

Figure 8-3. Average Concentrations of CO, PM and TVOC vs. Time for Energy loads of 400, 800, 1200 and 1900 watts.

temperature of the VAV/BPFS system decreased more than that of the VAV system, the zone damper of the VAV/BPFS system lets less air into the chamber than the corresponding VAV system. This indicates that the supply airflow rate of the VAV/BPFS system decreased more than that of the VAV system. According to resulting airflow balances, the pressure-drop from the outdoors to the inlet side of the fan is matched with the pressure-drop from the outlet side of the fan through the supply ducts to the chamber. The relative reduction of supply airflow rate of the VAV/BPFS system, therefore, results in a relative reduction of total system airflow rate and in a relative decrease of the outside air and in reducing energy consumption of the HVAC system.

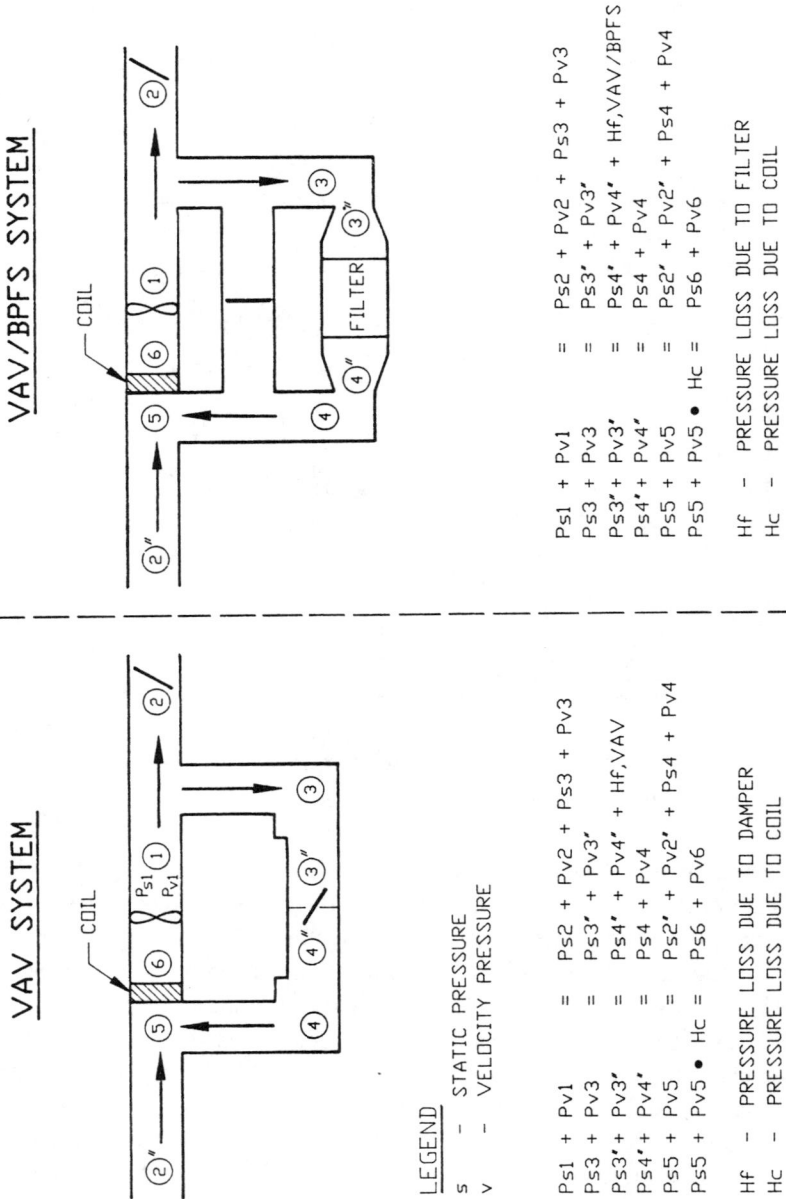

VAV/BPFS SYSTEM

$Ps1 + Pv1$ = $Ps2 + Pv2 + Ps3 + Pv3$
$Ps3 + Pv3$ = $Ps3' + Pv3'$
$Ps3' + Pv3'$ = $Ps4' + Pv4'' + Hf,VAV/BPFS$
$Ps4' + Pv4'$ = $Ps4 + Pv4$
$Ps5 + Pv5$ = $Ps2' + Pv2' + Ps4 + Pv4$
$Ps5 + Pv5 \bullet Hc$ = $Ps6 + Pv6$

Hf — PRESSURE LOSS DUE TO FILTER
Hc — PRESSURE LOSS DUE TO COIL

VAV SYSTEM

LEGEND
s — STATIC PRESSURE
v — VELOCITY PRESSURE

$Ps1 + Pv1$ = $Ps2 + Pv2 + Ps3 + Pv3$
$Ps3 + Pv3$ = $Ps3' + Pv3'$
$Ps3' + Pv3'$ = $Ps4' + Pv4'' + Hf,VAV$
$Ps4' + Pv4'$ = $Ps4 + Pv4$
$Ps5 + Pv5$ = $Ps2' + Pv2' + Ps4 + Pv4$
$Ps5 + Pv5 \bullet Hc$ = $Ps6 + Pv6$

Hf — PRESSURE LOSS DUE TO DAMPER
Hc — PRESSURE LOSS DUE TO COIL

Figure 8-4. Application of Bernoulli's Theorem to VAV and VAV/BPFS Systems.

Table 8-4. Total Energy Consumption for VAV and VAV/BPFS Systems.

ENERGY LOAD (Watts)	NUMBER OF EXPERIMENTS	TOTAL ENERGY CONSUMPTION (Btu/min.)	
		TEC $_{VAV}$	TEC $_{VAV/BPFS}$
400	27	239.3 ± 29.0 *	193.2 ± 20.1
800	27	266.5 ± 28.7	201.2 ± 25.3
1200	27	298.7 ± 23.2	242.3 ± 18.1
1900	27	320.9 ± 21.4	276.1 ± 13.2

* MEAN ± S.D.

The TEC increased with increasing energy loads. Energy consumption of the VAV system ranged from 239 Btu/min to 321 Btu/min, and that of the VAV/BPFS system ranged from 193 Btu/min to 276 Btu/min (refer to Table 8-4). Statistically, the null hypothesis (Ho : TEC_{VAV} = $TEC_{VAV/BPFS}$) of no difference in energy consumed by the two systems was rejected at α = 0.05 in favor of the energy consumed by the VAV/BPFS system.

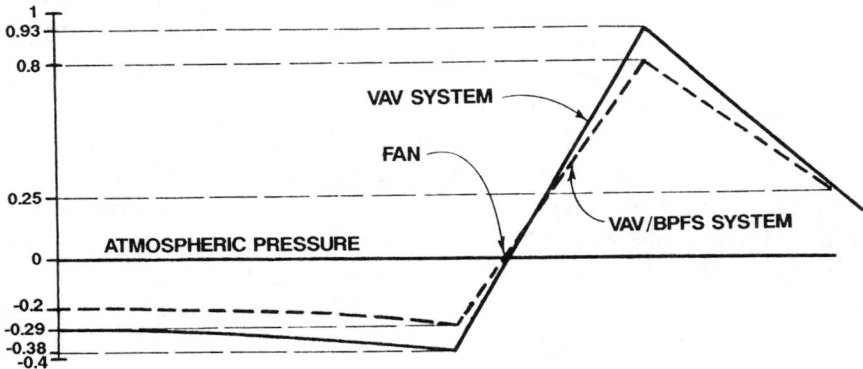

Figure 8-5. Pressure Distribution in Supply-Air Duct for VAV and VAV/BPFS Systems.

8.4   DISCUSSION AND CONCLUSIONS

Since infiltration is a chamber variable, the chamber infiltration rate did not change during the experiments. The outside supply airflow rates associated with the VAV/BPFS system were lower than those associated with the VAV system, but the v_{TER} of PM and TVOC of the VAV/BPFS system were higher than those of the VAV system. The increased v_{TER} of the VAV/BPFS system results from the effect of the carbon filter and particulate filter within the bypass filtration system.

Clearly, the greater the amount of fresh outside air introduced into an indoor environment, the better the IAQ. However, there is a penalty to be paid when the additional outside air must be conditioned (heated, cooled and dehumidified). The VAV/BPFS system reduces the need for reconditioning outside air and reduces the total system airflow rate. The VAV/BPFS system, therefore, reduces energy consumption in comparison to the VAV system.

Results obtained in this study show that the VAV/BPFS increases the total effective ventilation rate, and reduces indoor air contaminant concentrations without any penalty in energy consumption. The VAV/BPFS system decreases the indoor air contaminant concentrations at lower cost than those contaminant concentrations and energy consumption achieved with conventional VAV systems.

In summary, this study has concluded the following: (a) the VAV/BPFS system decreases outside supply airflow rate by 17% to 26% with respect to a conventional VAV system, (b) the performance of the VAV/BPFS system is more efficient than the VAV system in removing PM (compared to the v_{TER} of the conventional VAV system, the v_{TER} of the VAV/BPFS system increased by a relative amount of 53% to 62%), (c) the v_{TER} for TVOC of the VAV/BPFS system increased by a statistically significant amount of 15% to 22% in comparison to that of the VAV system, and (d) the TEC of the VAV/BPFS system is lower than that of the VAV system, (the energy savings of the VAV/BPFS system ranges from 8% to 14% in comparison to energy consumption of the VAV system).

Retrofitting the test chamber VAV system with the VAV/BPFS system was easy. The use of VAV/BPFS system is, therefore, recommended for buildings with a VAV system as well as for those under construction. Since these conclusions are drawn from the controlled chamber experiments, a series of "before and after retrofit experiments" in office buildings are required to further substantiate the cost-effectivity of the VAV/BPFS system in reducing indoor air contaminant concentrations.

8.5   REFERENCES

[1]McQuiston, F.C., and J.D. Parker, *Heating, Ventilating and Air Conditioning Analysis and Design*, 3rd ed., John Wiley & Sons, New York, NY., 1988.

[2]ASHRAE 1984, *System Handbook*, American Society of Heating, Refrigerating and Air-Conditioning Engineers, Inc., Atlanta, GA.

[3]Rickelton, D., and H.P. Becker, "Variable Air Volume," *ASHRAE Journal*, No. 9, pp. 31-55, 1972.

[4]Meckler, M., "Indoor Air Quality vs. Energy Efficiency," *Consulting-Specifying Engineer*, Vol. 4, pp. 82-88, 1988.

[5]ASHRAE 1985, *Fundamentals*, ASHRAE Handbook & Product Directory. American Society of Heating, Refrigerating and Air-Conditioning Engineers, Inc., Atlanta, GA.

[6]Moschandreas, D.J. et al., "Emission Rates from Unvented Gas Appliances," *Environment International*, Vol. 12, pp. 247-253, 1986.

9

Evaluation of a Liquid-Desiccant-Enhanced Heat-Pipe Air Preconditioner

Ahmad Pesaran, Ph.D.
Senior Engineer
and
Yves Parent, Ph.D.
Research Engineer
National Renewable Energy Laboratory
Golden, Colorado

Milton Meckler, P.E.
President, The Meckler Group
Encino, California
and
Davor Novosel
Project Manager, Gas Research Institute
Chicago, Illinois

9.1 INTRODUCTION

Because of increasing concerns about indoor air quality (IAQ) in buildings, ASHRAE Standard 62-1989 recommended the use of increased outdoor air ventilation rates (five cubic feet per minute [cfm] per person to 15 to 20 cfm per person). This additional outdoor air must be conditioned to the desired humidity and temperature before it is delivered to an occupied space. The conditioning of the ventilation air requires additional equipment and energy. When ASHRAE Standard 62-1989 is adopted, the cost of the equipment necessary to retrofit the existing commercial building systems was estimated to exceed $0.5 billion distributed over the next several years. New products are needed to condition the ventilation air in an energy-efficient and cost-effective manner. Furthermore, this new equipment must not use chlorofluorocarbons (CFCs), which are believed to contribute to depletion of the earth's ozone layer and whose production will be banned by the year 2000.

Conditioning of ventilation air requires temperature and humidity control. Furthermore, if the outdoor air is more polluted than indoor air, the ventilation air must also be cleaned. In the cooling season, the temperature of the outdoor air needs to be decreased. This can be accomplished by conventional vapor-compression units or exchanging heat with cool exhaust air using thermal recovery units such as heat wheels or heat-pipe air-to-air heat exchangers. In areas with medium-to-high outdoor humidity, dehumidification is required and heat exchange alone may not work. In this situation, desiccants can be used to add dehumidification capability to thermal recovery units[1]. In this chapter, we will explore the feasibility of using a liquid-desiccant-enhanced heat-pipe thermal recovery unit for preconditioning ventilation air. Additionally, the test procedure and the results will be presented. This study was funded by Gas Research Institute of Chicago, and the comparisons with enthalpy wheels were supported by National Renewable Energy Laboratory of Colorado.

A heat pipe is a closed heat-transfer device which relies on the vaporization and condensation of a working fluid to transport large quantities of heat energy at near-isothermal conditions. Figure 9-1 shows a simple heat-pipe. A bank of heat-pipe tubes assembled together forms a heat-pipe unit. Commercially available heat-pipe thermal energy recovery units can reclaim cooling energy during the summer and heating energy during the winter. The exchange takes place between the exhaust air and supply airstreams of a building (refer to Figure 9-2). The working section of the unit is composed of a bank of heat-pipe tubes with one end in the supply airstream and the other end in the exhaust airstream. In some commercial units, a cooler sink is provided in summer operation by spraying water on the exhaust-side to indirectly and evaporatively cool the supply air (refer to Figure 9-2).

Conventional heat-pipe thermal energy recovery units deal only with sensible cooling. In this system, a concentrated liquid desiccant is

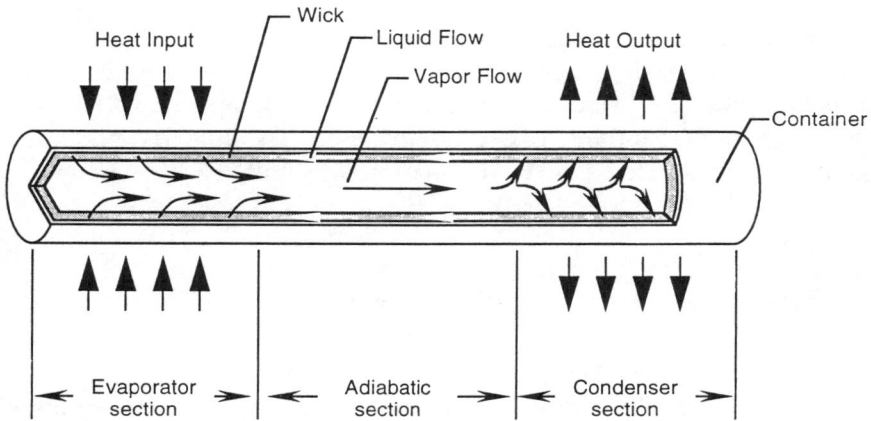

Figure 9-1. A Simple Heat-Pipe.

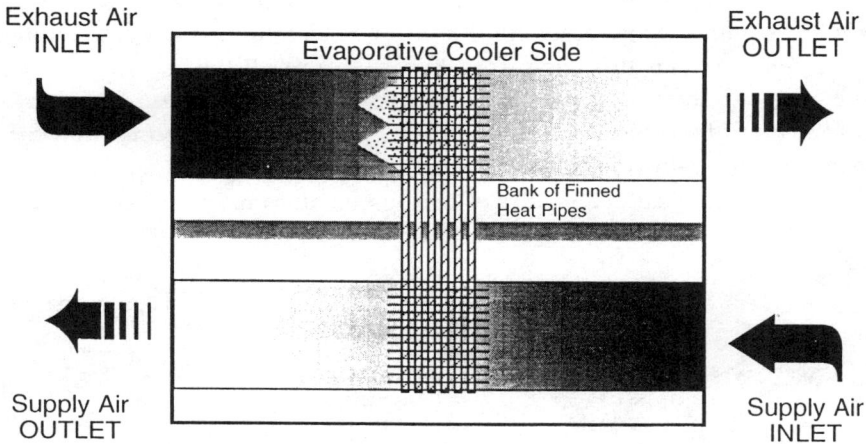

Figure 9-2. Heat-Pipe Thermal Recovery Unit with Indirect Evaporative Cooling.

sprayed on the heat-pipe fins on the supply-side to dehumidify the air (refer to Figure 9-3). The heat of absorption released during dehumidification is rejected to the exhaust-side through the heat-pipe tubes. The liquid desiccant becomes more effective because the heat of absorption is transferred away at the source, thus reducing the sorption temperature and positioning the operation in a more favorable portion of the desiccant/

moisture equilibrium map. The other advantage of this approach, expected to reduce component costs, is the integration of two conventional, separate processing steps (dehumidification and temperature change of the supply airstream) into one combined effect. The diluted liquid desiccant is reactivated for reuse by thermal energy input in a regenerator.

9.2   TESTING OF LIQUID-DESICCANT-ENHANCED HEAT-PIPE THERMAL RECOVERY UNIT.

9.2.1 System Description

A commercially available heat-pipe thermal recovery unit equipped with indirect evaporative cooling was purchased for the testing (refer to Figure 9-2). Water can be sprayed onto the contacting matrix from both the front (parallel to air flow) and top (perpendicular to air flow) of the matrix. At design conditions, the unit is capable of reclaiming 0.9 ton (3.52 kW) of cooling for summer conditions and 0.5 ton (1.76 kW) for winter conditions with an airflow rate of 600 cfm. Figure 9-2 shows a top view of the unit with the heat-pipes operating in a horizontal position. The working section of the test unit is composed of a bank of 48 aluminum heat-pipe tubes (5/8-inch in outer diameter) in a pattern of six staggered rows and an integral part of a section of corrugated aluminum fins set at 11 per

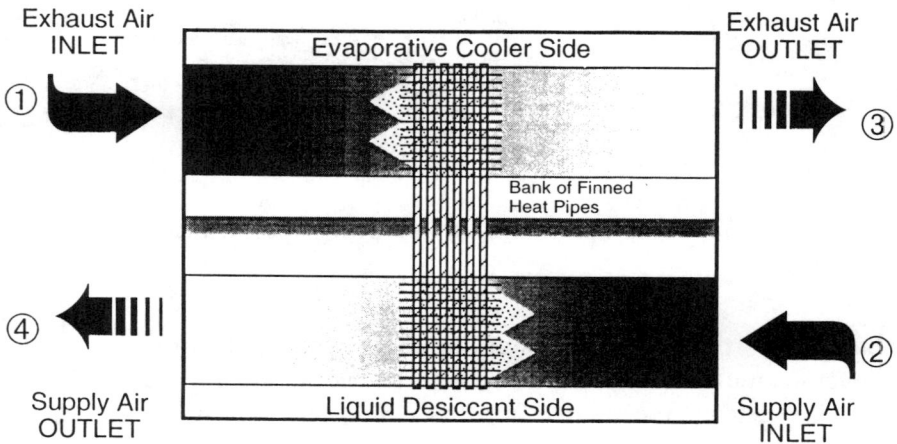

Figure 9-3. A Liquid-Desiccant-Enhanced Heat-Pipe Thermal Recovery Unit.

inch. The resulting exchange surface matrix on both ends of the heat-pipe structure fills the duct passage of the supply and exhaust airstreams, respectively. The dimensions of the contact volume thus formed are 12 inches by 11 inches with 8 inches in depth in the direction of flow. The true unobstructed width of the air duct is 10 inches.

Minor modifications were incorporated into the unit at the time of manufacture. For example, nozzles were installed to distribute the liquid desiccant onto the fin matrix on the supply-side and mist eliminators were installed to minimize the possibility of desiccant entrainment. The enclosure walls from both sides were made removable for easy access to all internal components. The standard tilt action for the heat-pipe component was eliminated thus, the conductance of the heat-pipe exchanger would remain fixed at whatever value was achievable in the set horizontal operating position. This would reduce the flexibility for winter operation, which was not of interest to us. Separate liquid distribution systems and sumps were provided to handle water on the exhaust-side and liquid desiccant on the supply-side. The water-side 1/12-hp (1 gallon per minute [gpm] pump was replaced with a larger one (1/8 hp) to allow flows up to 2.0 gpm. A 3/4-hp pump was installed to handle viscous liquid desiccants up to 2.0 gpm. A partition of closed-cell foam between the two sumps was added to minimize the exchange of moisture between these two parts of the system. These modifications to the standard unit are shown in Figure 9-3. To regenerate the liquid desiccant for the purpose of these experiments, a regeneration subsystem supplied by a major liquid desiccant system manufacturer was employed.

9.2.2 Experimental Test Setup

Figure 9-4 shows a simplified test setup for testing the liquid-desiccant-enhanced heat-pipe thermal recovery unit. The purpose of the test setup is to supply controlled airstreams at a desired flow rate, temperature and humidity to the unit. Furthermore, the dry-bulb temperature and the dew point temperature of the incoming and outgoing airstreams are measured and monitored to evaluate the performance of the test unit under such conditions. Copper-constantan thermocouple wires and chilled-mirror hygrometers are used to measure the air dry-bulb and dew point temperatures, respectively. Pressure-drops are measured using capacitance-type pressure transducers. The uncertainties of the measuring instruments are as follows: 0.5°F in dry-bulb temperature, 0.7°F in dew point temperature, and 0.03 inch of water in pressure-drop. The mass flow

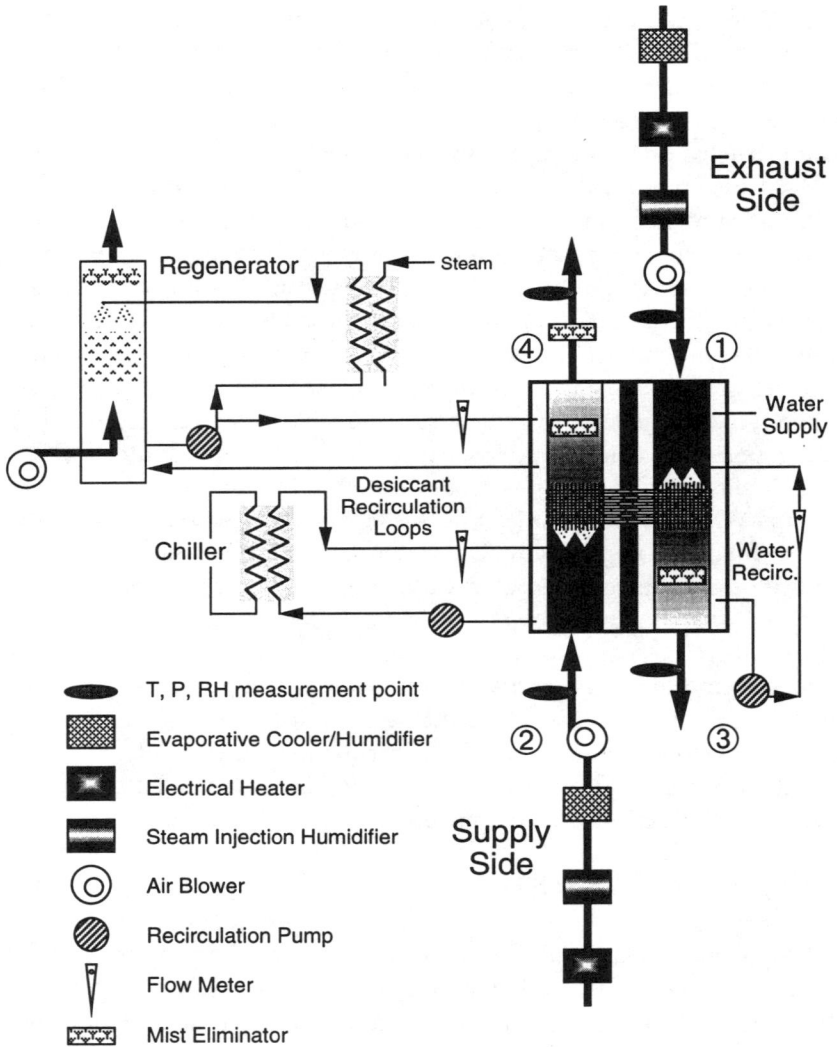

Figure 9-4. Test Setup for Evaluating Liquid-Desiccant-Enhanced Heat-Pipe Thermal Recovery Unit.

rate of each airstream is calculated (with an uncertainty of less than 3%) using the pressure-drop across the nozzles upstream of the heat-pipe and the absolute pressure is measured using a mercury barometer (with an uncertainty of less than 0.2%).

The humidity ratio at each measuring station is calculated (with an uncertainty of less than 3%) using the dew point temperature and the absolute pressure at that point. The desiccant concentration of selected samples is measured using a densimeter. The water mass flowrate for spray on the exhaust-side is measured using a rotameter, and the desiccant mass flowrate for spray on the supply-side is measured using a turbine flowmeter. A computer-controlled data acquisition system monitors and collects the data (air dry-bulb and dew point temperatures, pressure-drops, air mass flowrates, and liquid mass flowrates).

For the supply-side, room air is pulled into the preconditioning section by an electric blower. The incoming air temperature and moisture content can be controlled using an electric heater, direct steam injection, and direct contact evaporative cooling. Air contacts the liquid desiccant in the fin matrix of the unit. A coarse mist eliminator is located immediately after the fin pack to limit entrainment of desiccant droplets. An additional mist eliminator with finer mesh was built into the exhaust duct of the unit. The dry-bulb and dew point temperatures of the airstream are measured at the inlet to the unit. Once the air is through the unit and at the outlet, the airstream dry-bulb and dew point temperatures are measured once again. The air is then vented to the outside of the building.

On the supply-side, the unit interacts with two semi-closed circulation loops of liquid desiccant. The loop having one pump draws liquid desiccant from the bottom of the unit's sump and returns a portion of this volume directly to the sump to ensure mixing of the sump volume, while the remaining circulating volume is sent through a cooled heat-exchanger and onto the spray heads of the dehumidifier section of the unit. The flow rate to the spray manifold is controlled. Cooling in the heat-exchanger is ensured by circulating cold water from a chiller. The second desiccant circulating loop interconnects the unit to the regenerator. Desiccant overflows from the unit's sump moves by gravity toward the regenerator sump.

At the regenerator, a pump draws liquid from its sump, sends a small amount of liquid back to the unit's sump at a controlled flow rate and sends the remainder of the flow volume to a steam-heated plate heat-exchanger and onto the spray head of the regenerator. Regeneration is accomplished by counter-current direct contact of room air (up flow) with hot desiccant solution sprayed onto a packed bed of commercial column packing. The liquid desiccant is heated in a fin-and-plate heat-exchanger using steam. Control of the heat input is based on liquid level in the sump.

Air-handling on the exhaust-side is similar to that of the supply-side. Inside air, outside air or a blend of both is drawn in by a blower. The incoming air temperature and moisture content can be adjusted and controlled as described above. The dry-bulb and dew point temperatures of the airstream are also monitored at the inlet to the unit using the type-K thermocouples and chilled-mirror hygrometers, respectively. Once the air is through the unit and at the outlet, the airstream dry-bulb and dew point temperatures are measured once again. The air is then vented to the outside of the building.

On the exhaust-side, only one circulation loop ensures flow and distribution of the water for the evaporative cooler operation. The sump is automatically replenished with water by using a float valve. A small submersible pump draws water from the sump, feeds a direct-return line for mixing purposes, and also feeds the spray manifold of the evaporative cooler. The water spray flowrate is monitored and controlled.

9.3   TEST PROCEDURE

Three types of operating conditions were considered (refer to Table 9-1): (a) operating without spraying any desiccant on the supply-side, (b) operating with water spraying on the exhaust-side, and (c) operating without water spraying on the exhaust-side. The data set established the baseline and allowed comparison to the conventional design specifications. Tests LiCl.1 through LiCl.8 and TEG.1 through TEG.6 in Table 9-1 represent such conditions. Second, the use of a lithium chloride (LiCl) solution as a desiccant was addressed. In this set, both operation with/ without water on the exhaust-side were considered. Variations in inlet temperature and moisture content on supply-and exhaust-sides were investigated. Tests LiCl.9 through LiCl.16 deal with these operating conditions. Finally, the tests were run with triethylene glycol (TEG) solutions. Tests TEG.7 through TEG.16 cover these particular operating conditions.

Another purpose of the test procedure was to observe the operation of the various components of the system to identify possible issues and recommend improvements. The commercial unit was not specifically designed to handle viscous liquid desiccants such as LiCl and TEG in the

Table 9-1. Test Conditions and Resulting Heat-Pipe Transfer Loads.

Test#	SUPPLY AIR Inlet moisture content		EXHAUST AIR Inlet moisture content		LIQUID FLOW RATE[d] water	desiccant	ENERGY TRANSFER heat pipe[c]
ID	lb/lb	[%RH][a]	lb/lb	[%RH][b]	[gpm]	[gpm]	[tons cooling]
LiC1.1	.0043	19 @ 75°F	.0043	19	0.0	0.0	n/a
LiC1.2	.0046	11	.0045	20	0.0	0.0	0.41
LiC1.3	.0045	10	.0046	20	2.0	0.0	1.27
LiC1.4	.0044	10	.0044	20	2.5	0.0	1.28
LiC1.5	.0120	28	.0051	23	2.0	0.0	1.26
LiC1.6	.0201	46	.0057	25	2.0	0.0	1.30
LiC1.7	.0209	48	.0066	29	2.0	0.0	1.41
LiC1.8	.0194	48	.0093	41	2.0	0.0	1.33
LiC1.9	.0134	29	.0060	26	0.0	1.8	0.35
LiC1.10	.0125	29	.0059	26	2.0	1.8	1.51
LiC1.11	.0184	42	.0062	27	2.0	1.8	1.61
LiC1.12	.0188	43	.0068	30	2.0	1.8	1.73
LiC1.13	.0182	42	.0082	36	2.0	1.8	1.75
LiC1.14	.0183	42	.0110	48	2.0	1.8	1.51
LiC1.15	.0113	26	.0059	26	0.0	1.8	0.33
LiC1.16	.0184	43	.0116	51	2.0	1.0	1.41
TEG.1	.0180	42	.0115	50	0.0	0.0	0.42
TEG.2	.0186	42	.0116	51	1.0	0.0	0.98
TEG.3	.0178	41	.0116	49	2.0	0.0	1.07
TEG.4	.0181	42	.0076	34	0.0	0.0	0.46
TEG.5	.0184	42	.0080	35	1.0	0.0	0.95
TEG.6	.0186	43	.0083	36	2.0	0.0	1.28
TEG.7	.0179	41	.0117	51	0.0	1.0	0.55
TEG.8	.0176	41	.0116	50	1.0	1.0	0.83
TEG.9	.0182	42	.0115	51	2.0	1.0	1.40
TEG.10	.0183	42	.0116	51	2.0	1.5	1.37
TEG.11	.0180	42	.0086	38	0.0	1.0	0.48
TEG.12	.0177	42	.0089	40	1.0	1.0	0.97
TEG.13	.0173	40	.0091	40	2.0	1.0	1.32
TEG.14	.0178	41	.0118	52	2.0	1.0	1.43
TEG.15	.0181	42	.0121	54	2.0	1.5	1.39
TEG.16	.0179	42	.0115	51	2.0	1.0	1.34

air flow rates are 628 acfm
a: at 95°F (35°C), unless otherwise specified
c: energy transferred by the heat pipe bundle
b: at 75°F (23.9°C), unless otherwise specified
d: 0.0 means no water or desiccant sprayed on the exhaust or supply sides, respectively

most optimum manner such as fin spacing and flow distribution to allow maximum liquid coverage.

Air inlet flowrate could be varied but most tests were conducted at approximately 630 acfm. Similarly, the flows of both liquid desiccant and water to the contacting sections were set by visual inspection of adequate spray pattern at the nozzles. These conditions correspond to flow rates of 2.0 gpm for water and 1.8 gpm for LiCl. In the case of TEG, the viscosity and pressure-drop in the nozzle line limited the operation to the flow rate of 1.0 gpm to 1.5 gpm. The return flowrate of the desiccant from the regenerator to the unit sump could be adjusted and maintained at flow rates up to 2.0 gpm.

Data were obtained for 32 different tests as shown in Table 9-1[3]. Table 9-2 presents a selection of parameters that were kept essentially constant throughout these tests. Inlet air conditions for the supply- and exhaust-sides are selected near the Air Conditioning and Refrigeration Institute (ARI) conditions for the Denver elevation. The essential data describing differences between all the tests completed are presented in Table 9-1. A quick overview of performance is also given in this table by tabulating the energy transfer between the supply-side and the exhaust-side as attributable to the heat-pipe bundle. These values are expressed in equivalent tons of cooling.

Table 9-2. A Sample of Nominal Operating Conditions for Tests Conducted.

	SUPPLY SIDE	EXHAUST SIDE
INLET Air Temperature	95°F (35°C)	75°F (24°C)
INLET Air Mass Rate	37.4 lb/min (0.283 kg/s)	37.4 lb/min (0.283 kg/s)
Water Spray Rate*		2 gpm (456 L/s)
Desiccant Spray Rate*	1-1.8 gpm (228-408 L/s)	
Desiccant Return Rate*	1.4-1.8 gpm (318-408 L/s)	
Desiccant Concentration*	37wt% LiCl, 95wt% TEG	
Chiller Temperature	50°F (10°C)	

* if in use

9.4   RESULTS AND DISCUSSION

9.4.1 Air Flow Pressure-Drop Across Unit

Measurements of pressure-drop across the unit (including heat-pipe bundle, coarse mist eliminator, and inlet and outlet duct transition sections) with and without a liquid flow yielded the following results for air flows of 628 acfm.

- No liquid flows, same airflow rates on both sides, same temperature and same humidity,

- Supply-side: ΔP= 0.79 inch of water
- Exhaust-side: ΔP= 0.85 inch of water

- With the uncertainty of less than 0.03 inch of water, these values are comparable and suggest the system is relatively well balanced in terms of resistance to air flow.

- The presence of liquid flow across the fins resulted in an increase of ΔP, as evidenced in all water tests. For example,

- Supply-side: ΔP= 1.20 inches of water at 1.8 gpm LiCl (37 weight[wt]-%)

- Supply-side: ΔP= 1.17 inches of water at 1.5 gpm TEG 95 weight-%

- Exhaust-side: ΔP= 1.08 inches of water at 2.0 gpm water

9.4.2 Overall Performance Without a Desiccant

In operating without a desiccant, the heat-pipe performed at least as well as rated by the manufacturer. Tests LiCl.1 through LiCl.8 and TEG.1 through TEG.6 were all performed without a desiccant. In the evaporative cooling mode, the heat-pipe delivered between 0.9 and 1.2 tons of cooling to the incoming supply stream. The improved performance (over manufacturer's specification) can be associated with the increased water flowrate at the evaporative cooler and to the reduced air flow on both sides of the heat-pipe. This performance was apparently unaffected over a relatively wide range of moisture contents for both the supply- and exhaust-sides.

Comparing test conditions of LiCl.7 and TEG.6, one would expect a higher energy transfer rate with Test TEG.6 which has drier air on the inlet of the exhaust-side. However, the energy transfer rate for Test TEG.6 is about 0.05 ton less. Such discrepancies in energy transfer are likely associated with the unsteady behavior of the system during the data acquisition for these sets. Indeed, although the desiccant was not being sprayed on the dehumidifier side, it was still being recirculated between the regenerator and the sump. Some variation in the temperature of the liquid desiccant in the sump may have contributed to energy transfers to either side of the heat-pipe.

In the evaporative cooling mode of operation, the heat-pipe performs as a relatively efficient sensible heat-exchanger for the supply-side. Figure 9-5 shows a typical example of the psychrometric process for this mode of operation. Note that points 1 and 3 on the psychrometric chart apply to the exhaust-side process and correspond to inlet and outlet as indicated in Figure 9-3, respectively. Similarly, points 2 and 4 apply to the supply-side and correspond to inlet and outlet, respectively.

9.4.3 Overall Performance With a Desiccant

The operation with desiccants included LiCl and TEG tests. The results of each group of tests are discussed below.

9.4.3.1 LiCl Tests

Tests LiCl.9 through LiCl.16 were conducted with LiCl as a desiccant. Two of these, Tests LiCl.9 and LiCl.15, were carried out without water (i.e., the evaporative cooler was turned off). As discussed earlier, energy transfer is much lower in such a case. The strong effect of evaporative cooling can again be observed between Tests LiCl.9 and LiCl.10, where the additional cooling enhances the desiccating power of the liquid from 17 grains/lb to 21 grains/lb.

Note that two factors can limit accuracy of the energy balances: (a) large masses of water and desiccant in the respective sumps and (b) the unsteady behavior of the system such as the off/on cycle of the regenerator during data acquisition. These factors may lead to some discrepancies in the heat transfer rates and may increase uncertainty (about + 15%). For example, the observed behavior from Tests LiCl.11 through LiCl.13 was not expected; it was expected a steady decrease with an increase in the inlet exhaust-air humidity ratio.

Tests LiCl.12 through LiCl.14 constitute the most representative inlet conditions for the system. Performance is on the order of 1.7 tons for Tests LiCl.12 and LiCl.13, in which the exhaust inlet moisture content increases. Performance falls off to 1.5 tons of cooling as the exhaust inlet moisture content increases to almost 50% relative humidity (RH), thus reflecting the

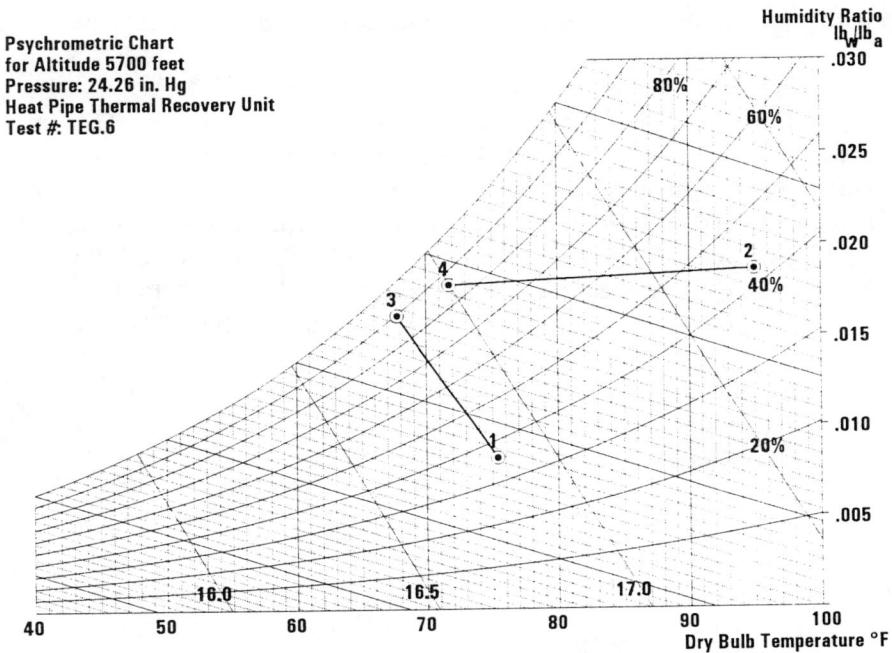

Psychrometric Chart
for Altitude 5700 feet
Pressure: 24.26 in. Hg
Heat Pipe Thermal Recovery Unit
Test #: TEG.6

State Point Data

#	Dry Bulb °Fdb	Wet Bulb °Fwb	Dew Point °Fdp	Relative Humidity %RH	Humidity Ratio lbw/lba	Specific Volume lb/cu ft	Enthalpy Btu/lb
1	75.40	57.32	46.45	35.77	0.00825	16.86	27.12
2	95.00	75.01	68.53	42.34	0.01858	17.77	43.29
3	67.80	65.50	64.56	89.44	0.01614	16.83	33.88
4	71.80	68.52	67.31	85.81	0.01780	17.00	36.68

Figure 9-5. Psychrometric Chart Showing Heat-Pipe Operation without a Desiccant for Test TEG.6.

decrease in evaporative cooling capacity. Test LiCl.16 is the operation of the system for several hours during which conditions were maintained constant to observe the steady-state response. The performance is slightly lower than the previous tests, most likely because of operation at a reduced desiccant flow rate (1.0 gpm). The specifics of this test are shown in Figure 9-6. Overall, the heat-pipe bundle transfers 30% to 40% more heat when operating with LiCl. The nominal concentration for the LiCl solution in these tests was 37 wt-% LiCl. LiCl solutions can be corrosive and incompatible with ordinary metals. Coating or replacing these metals with specialty metals or plastics may increase the cost of the system.

9.4.3.2 TEG Tests

Tests TEG.7 through TEG.16 represent operations using TEG as a desiccant. Two of these tests, TEG.7 and TEG.11, were carried out without water (i.e., the evaporative cooler was turned off). They show that the resulting energy transfer is much lower. Comparing the results of the tests conducted with TEG (Tests TEG.9 and TEG.10, and TEG.13 through TEG.16) shows that heat-pipe bundles transfer about 20% to 30% more heat when operating with TEG as a desiccant. For the TEG tests, the

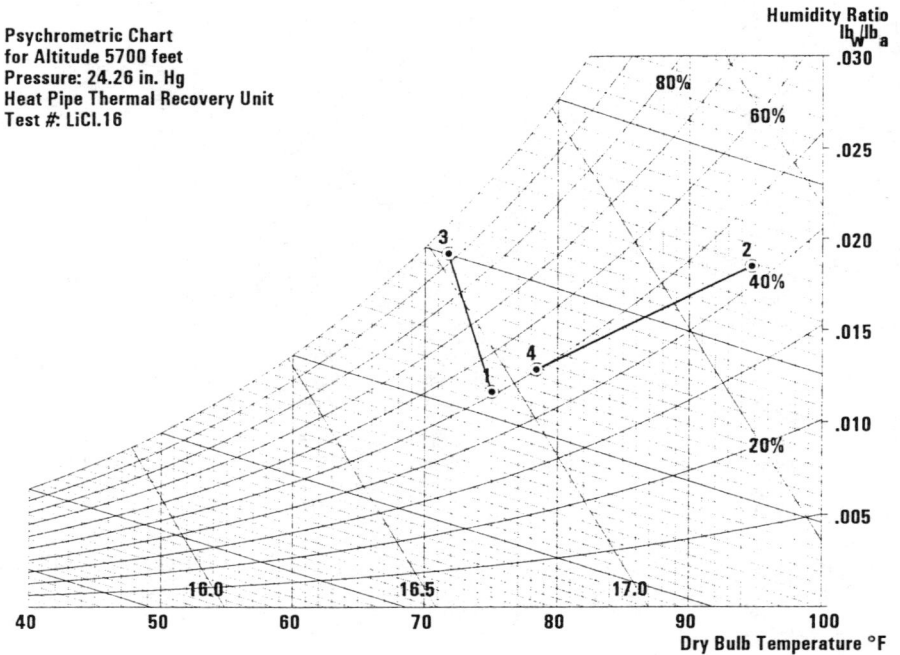

State Point Data

#	Dry Bulb °Fdb	Wet Bulb °Fwb	Dew Point °Fdp	Relative Humidity %RH	Humidity Ratio lbw/lba	Specific Volume lb/cu ft	Enthalpy Btu/lb
1	75.00	61.91	55.53	50.83	0.01163	16.94	30.72
2	94.80	74.85	68.35	42.34	0.01846	17.76	43.11
3	71.80	70.06	69.45	92.34	0.01919	17.04	38.20
4	78.40	64.49	58.26	50.10	0.01285	17.08	32.90

Figure 9-6. Psychrometric Chart Showing Heat-Pipe Operation with LiCl for Test LiCl.16.

nominal concentration of the solution was 95 wt-% TEG. Test TEG.16 operated for several hours to observe the steady-state response. The performance, in terms of latent load on the desiccant, is slightly lower than in previous tests. Over the period of several hours of operation for Test TEG.16, the regenerator loop operation stabilized at a slightly higher moisture content in the solution. Specifics for this test are presented in Figure 9-7.

All tests using TEG show lower heat-pipe duty (about 10% on the average) than in the case of the LiCl tests. The two long-run tests (LiCl.16 and TEG.16), which reduce unsteady behaviors through averaging, show

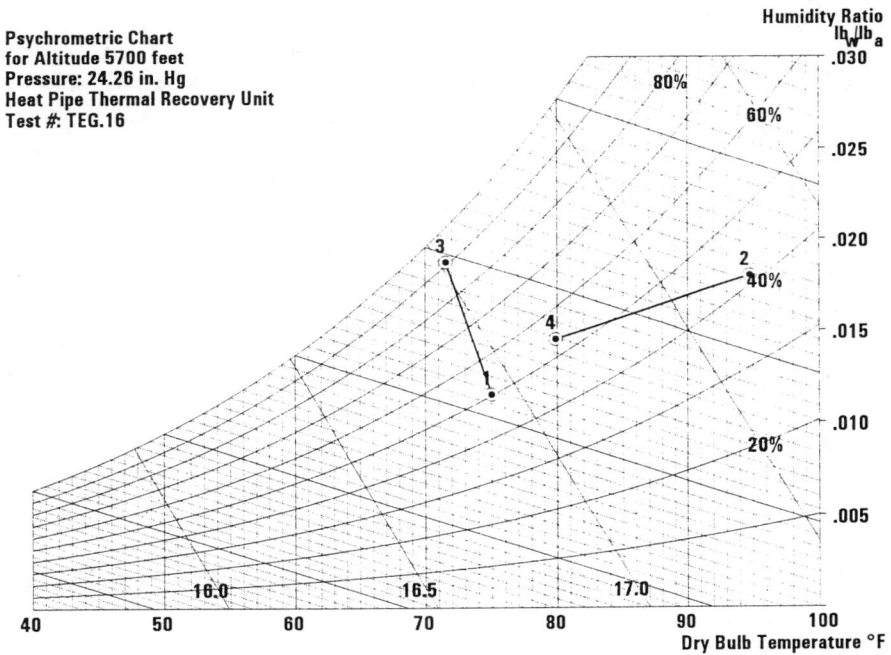

Psychrometric Chart
for Altitude 5700 feet
Pressure: 24.26 in. Hg
Heat Pipe Thermal Recovery Unit
Test #: TEG.16

State Point Data

#	Dry Bulb °Fdb	Wet Bulb °Fwb	Dew Point °Fdp	Relative Humidity %RH	Humidity Ratio lbw/lba	Specific Volume lb/cu ft	Enthalpy Btu/lb
1	75.00	61.73	55.22	50.25	0.01149	16.94	30.57
2	94.80	74.35	67.56	41.20	0.01795	17.74	42.55
3	71.60	69.50	68.75	90.78	0.01873	17.02	37.64
4	79.90	66.92	61.60	53.72	0.01451	17.17	35.09

Figure 9-7. Psychrometric Chart Showing Heat-Pipe Operation with TEG for Test TEG.16.

lower heat duty for TEG. Figure 9-8 shows a typical comparison of the heat-pipe bundle performance for three cases (with LiCl and TEG, and without a desiccant) under similar conditions. The cooling performance of the heat-pipe with TEG at 1.4 tons is about 10% less than that of the heat-pipe with LiCl at 1.51 tons, and about 30% more than that of the heat-pipe without any desiccant. The lower heat transfer rate and associated lower moisture removal rate by TEG can be attributed to less efficient contact between the desiccant and supply airstream. Higher viscosity of TEG, leading to poor distribution of the liquid in the contacting fin matrix, can be a cause of less efficient exchange. In addition, the specific sorption energy of water by TEG is less than by LiCl. Attempts at improving mass transfer in the dehumidifier section by increasing the desiccant flowrate (Test TEG.9 compared with Tests TEG.10 and TEG.15) did not yield any improvement. These observations confirm the inadequate distribution of liquids and inadequate design of the contacting section (tubes/fins spacings) for liquids significantly different than water.

A potential advantage of use of TEG solutions is their capability of cleaning air by removing volatile organic compounds (VOCs)[2]. Although TEG has a very low volatility, it may be carried over by the supply air and later condensed on cold surfaces in a building; engineering solutions exist to prevent this. Organic liquids with higher molecular weight such as polyethylene glycols can also be used instead.

9.4.3.3. Both Desiccants

Because of poor mass transfer, neither desiccant exhibits its maximum dehumidification capacity. The supply-side outlet air could have been much drier if it was approaching equilibrium. In our tests, LiCl and TEG exhibited only 50% and 40% of their dehumidification potential, respectively. By improving mass transfer rate (e.g., through improved spraying and flow distribution, improved fin configuration and spacing, and increased contact area), it is feasible to achieve 85% of the dehumidification potential of the desiccant, which can result in drier air and higher latent-load removal. Improving the mass transfer rate will be a challenge for further research and development.

For each experiment, the heat added by the regenerator was estimated, and the heat removed by the chiller was approximated based on mass flow rates and temperature changes. For example, for Test LiCl.16, the heat added by the regenerator was 4750 Btu/h and the heat removed by the chiller was 1.0 ton. Note that the heat-pipe energy transfer rates

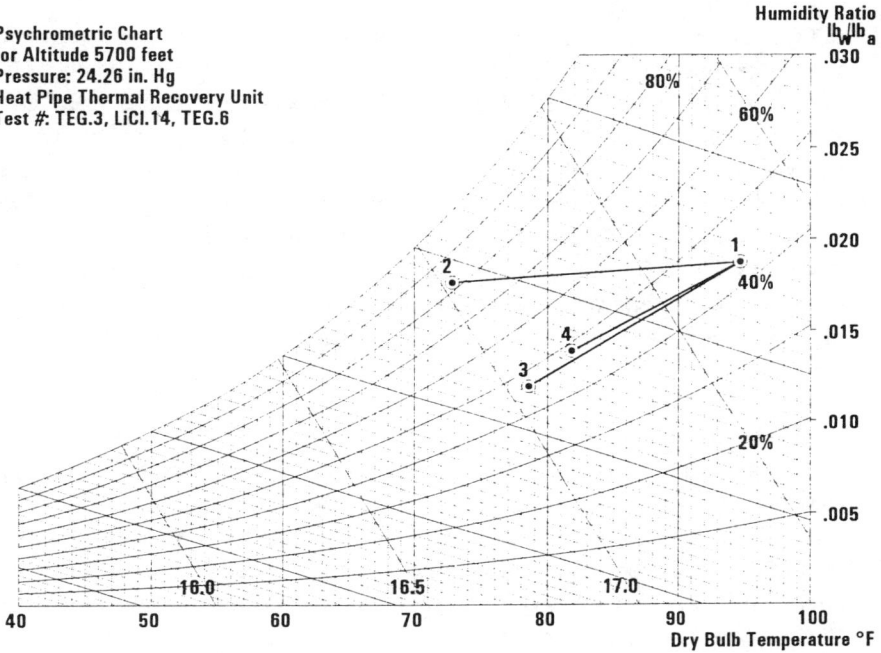

Psychrometric Chart
for Altitude 5700 feet
Pressure: 24.26 in. Hg
Heat Pipe Thermal Recovery Unit
Test #: TEG.3, LiCl.14, TEG.6

State Point Data

#	Dry Bulb °Fdb	Wet Bulb °Fwb	Dew Point °Fdp	Relative Humidity %RH	Humidity Ratio lbw/lba	Specific Volume lb/cu ft	Enthalpy Btu/lb
1	94.74	75.09	68.76	43.01	0.01873	17.76	43.39
2	72.86	68.62	67.05	82.07	0.01764	17.03	36.76
3	78.60	63.42	56.29	46.38	0.01196	17.06	31.97
4	81.86	66.71	60.37	48.25	0.01388	17.22	34.87

Point #	Description
1	Entering Supply Air
2	Leaving Supply Air from Tested Heat Pipe without Desiccant
3	Leaving Supply Air from Tested Heat Pipe with LiCl
4	Leaving Supply Air from Tested Heat Pipe with TEG

Figure 9-8. Comparison of Heat-Pipe Bundle Performance with LiCl and TEG, and without a Desiccant.

shown in Table 9-1 are calculated by excluding the contribution of cooling by the chiller. Because the desiccant volume and the size of the regenerator in the system was larger than one would actually use for a practical unit, no attempt was made to estimate any coefficient of performance

(COP) for the system. The earlier analysis[3] indicates that for a regeneration latent COP of 0.55, 1.0 and 1.6, the thermal COP of the tested TEG-enhanced heat-pipe is 0.93, 2.06 and 3.74, respectively. The latent COP is defined as the latent load removed divided by the thermal energy input to the regenerator. The thermal COP can be defined as cooling capacity (sensible and latent) removed divided by the thermal energy input. Note that both of these COPs exclude any electrical energy for pumps and fans. A latent COP of 0.55 is achievable with the existing standard regeneration equipment. Regenerators with latent COPs of 1.6 need to be developed and may require a more sophisticated regeneration approach such as staging.

9.4.4 Comparison with Enthalpy Wheels

As discussed earlier, the liquid-desiccant-enhanced heat-pipe thermal recovery unit can be used for preconditioning ventilation (outside) air. Using a preconditioner to treat outside air rather than using the main vapor compression/air conditioner (VC/AC) system can save energy costs. There are a number of units that can be used for treating outside air. Enthalpy wheels, considered among efficient commercially available heat-recovery units, can be used for treating outside air[4]. The tested LiCl-enhanced heat-pipe unit, in its present configuration and without further improvements in gas/liquid contacting operation, performed in a comparable manner to a high performance (80% effective) commercially available rotating enthalpy wheel.

Energy costs of a VC/AC system was compared with and without enthalpy wheels or TEG-enhanced heat-pipe preconditioners[3]. Because either the enthalpy wheel or the TEG-enhanced heat-pipe is used as a preconditioner, their ultimate performance comparison should be made when they are integrated with a VC/AC treating the recirculated air as shown in Figure 9-9. Energy needs for conditioning 1000 cfm of outside air in Denver from the ARI design air condition to a supply design air condition of 55°F and 80% RH were compared[3]. Three options to provide the 5.82 tons of cooling were compared: a VC/AC, an 80%-efficient enthalpy wheel integrated with a VC/AC, and an 85%-efficient TEG-enhanced heat-pipe integrated with a VC/AC. The latent COP for TEG regeneration was assumed to be 1.6. The electric COP of the VC/AC was assumed to be 3.0 in all three options. The electric energy consumption for the preconditioners was estimated based on parasitic power needs. The parasitic loss for enthalpy wheels was based on air-side pressure-drops from manufacturer's literature[4]. The parasitic loss for the desiccant-enhanced heat-pipe was based on: (a) measured pressure-drops on air-sides and (b) the power needs to pump and distribute water and liquid desiccant[3].

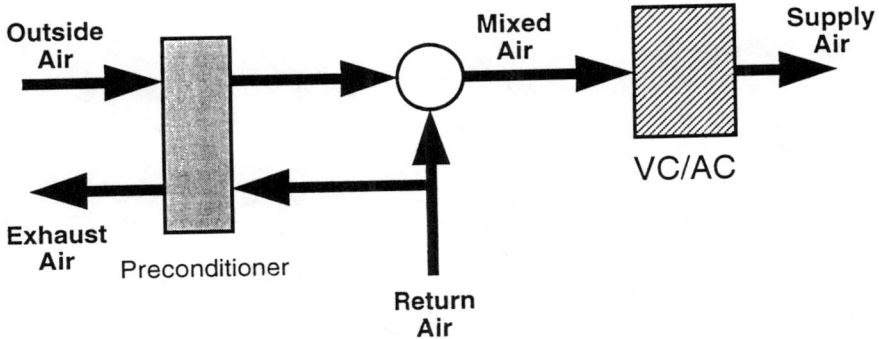

Figure 9-9. An Integrated Preconditioner with a Vapor-Compression/ Air Conditioner System.

Figure 9-10 compares hourly operating energy costs (in cents per hour) for the three options discussed above. Natural gas, with a burning efficiency of 95%, was used for providing regeneration energy to the TEG heat-pipe. Clearly, the operating energy costs for the VC/AC only option is higher than the other two preconditioner options. The operating energy cost of the VC/AC and the TEG-enhanced heat-pipe option is less than that of VC/AC and enthalpy wheel at low gas prices and moderate-to-high electricity prices. The liquid-desiccant-enhanced heat-pipe heat recovery unit can offer energy cost savings over the enthalpy wheel. Although this unit may be more complex than an enthalpy wheel, it has the advantage of removing contaminants from the air[2].

9.5   CONCLUSIONS AND RECOMMENDATIONS

The feasibility of enhancing the dehumidification capability of a heat-pipe thermal recovery unit was tested by contacting with a liquid-desiccant for preconditioning supply air. The heat-pipe unit incorporated indirect evaporative cooling. Data showed that the heat-pipe with TEG on the supply-side provided about 30% more cooling capacity than the heat-pipe without a desiccant. The cooling performance of the heat-pipe with

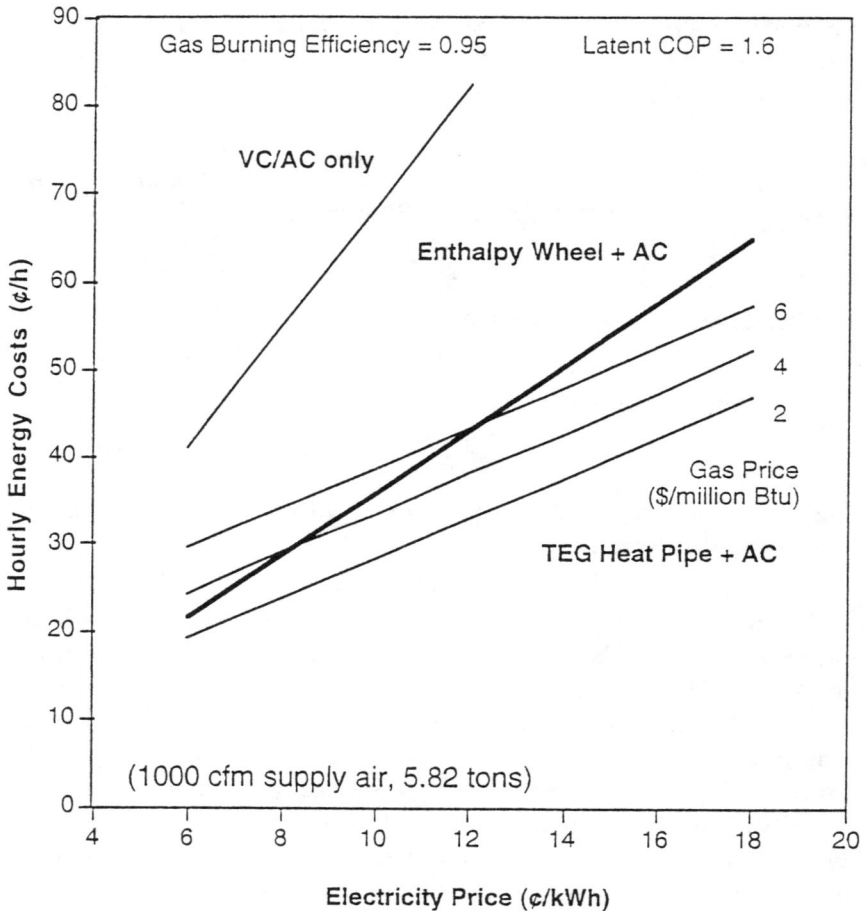

Figure 9-10. Hourly Energy Cost Comparison of Three Options For Preconditioning 1000 cfm of Outside Air in Denver, CO (ARI Design Conditions).

LiCl was even better, about 40% more than the systems without a desiccant. The liquid-desiccant-enhanced heat-pipe transfers both latent and sensible heat between the streams in a very compact and mechanically simple system. No mechanical refrigeration using CFCs is involved, and the system easily lends itself to retrofit situations where increases in ventilation rates are anticipated.

Based on equilibrium considerations it was found that in the conducted tests only about 40% to 50% of the dehumidification potential of liquid desiccants was used. This was attributed to poor mass transfer rates because of poor distribution of the desiccant in the flow channels and insufficient contact area between heat-pipe surfaces, the liquid desiccant, and the airstream. Further research and development is needed to improve the mass transfer rate by improving flow distribution and fin spacing, increasing the contact area, thus achieving higher efficiencies of up to 85% the dehumidification potential.

The system showed comparable performance to an 80%-effective enthalpy wheel. With further improvements in design of the dehumidifier section, the latent load handled can be doubled and the performance of the TEG heat-pipe can exceed the performance of an enthalpy wheel. Estimates indicated that at low-to-moderate gas prices and moderate-to-high electricity prices, the desiccant-enhanced heat-pipe integrated with a vapor compression air-conditioner is less expensive to operate than an enthalpy wheel integrated with an air-conditioner when the latent (or regeneration) COP is above 1.4. The analysis did not consider the potential use of TEG for removing contaminants such as VOCs.

The data generated from the experiment clearly show that the integration of direct-contact dehumidification by means of a liquid-desiccant (liquid-desiccant-enhanced heat-pipe) is feasible. Both sensible and latent loads can be removed without the use of CFCs or hydrochlorofluorocarbons (HCFC). The liquid-desiccant-enhanced heat-pipe can be a viable heating, ventilating and air-conditioning (HVAC) component for preconditioning outside ventilation air.

Based on the promising results obtained in this study, further development of the liquid-desiccant-enhanced heat-pipe preconditioner for treating ventilation air is recommended. Efforts to improve the dehumidification capability of the desiccant heat-pipe up to 85% of equilibrium dehumidification potential are recommended. This might be achieved by improving the mass transfer process by improving the desiccant flow distribution between heat-pipe fin channels and also extending the finned matrix in the direction of air flow.

Engineering solutions to avoid potential carryover of TEG in the supply air because of its low volatility need to be tested and evaluated. Liquid desiccants with higher molecular weights such as polyethylene glycols should be evaluated for this application. Another important component that was not studied here is the desiccant regenerator. To compete successfully with other preconditioners, the liquid-desiccant-enhanced heat-pipe unit requires an efficient regenerator. Investigations for designing efficient regenerators such as multistage or vapor-compression distillation devices are recommended.

Improved liquid-desiccant-enhanced heat-pipes and regenerators need to be designed, fabricated and tested in the laboratory. Compact designs for the liquid-desiccant-enhanced heat-pipe as an add-on component to the existing packaged units need to be fabricated and field-tested.

9.6   REFERENCES

[1]Meckler, M., "Integrated Desiccant Cold Air Distribution System," *ASHRAE Transactions*, Vol. 95, Pt. 2, 1989.

[2]Hines, A.L., and T.K. Ghosh, "Investigation of Co-Sorption of Gas and Vapors as a Means to Enhance Indoor Air Quality - Phase 2: Air Dehumidification and Removal of Indoor Pollutants by Liquid Desiccants," GRI-92/0157.3, Gas Research Institute, Chicago, IL., 1992.

[3]Parent, Y., A.A. Pesaran, and M. Meckler, "Evaluation of Dehumidifiers with Polymeric Liquid Desiccants," GRI-93/0194, Gas Research Institute, Chicago, IL., December 1993.

[4]Semco Manufacturing Inc. *Exclusive Design and Selection Manual*, Brochure ER-503, Columbia, Missouri, 1989.

10

Co-Adsorption of Indoor Air Contaminants By Adsorbents

Tushar K. Ghosh, Ph.D.
University of Missouri-Columbia,
Columbia, Missouri
and
Anthony L. Hines, Ph.D., P.E.
Honda of America Mfg., Inc.
Marysville, Ohio

10.1 INTRODUCTION

Porous materials, such as silica gel, molecular sieves and activated carbons, have a unique capability of selectively adsorbing various chemical compounds from air or an inert gas stream. This property has been exploited in designing various air-cleaning devices. Over the past several years, solid desiccant-based air conditioning systems have found increased applications as humidity-control devices for non-industrial structures, such as schools, homes, hospitals and commercial buildings. These systems offer the added advantage of enhancing indoor air quality (IAQ). Solid desiccants that include silica gel and molecular sieves can simultaneously remove moisture and contaminants from indoor air by co-sorption. If the sole objective is to clean the air, other types of adsorbents such as activated carbons, can be used. Activated carbons are also capable of co-adsorbing a number of organic compounds in the presence of moisture. However, the mechanisms for co-sorption are different for different types of adsorbents.

Indoor environments contain a variety of contaminants, including volatile organic compounds (VOCs), inorganic gases that result from combustion byproducts, and bioaerosols. More than 300 contaminants have been identified indoors, but not all of them are present simultaneously in an indoor environment. At any given time, approximately 50 to 60 contaminants are present indoors, and all of them will compete for adsorption sites on an adsorbent. Solid desiccants generally have a greater affinity for water than for the organic or other inorganic compounds, whereas activated carbons have the greatest capacity for most of the organic compounds than any solid desiccant.

In a mixture, the adsorption characteristics of both water and contaminants are different from their pure component adsorption behavior. During adsorption of water and contaminants from their mixtures on silica gel or molecular sieves, water can displace the weakly adsorbed components (contaminants) from the pores. The concentration of the weakly adsorbed compound in the outlet stream becomes higher than its inlet concentration within a few minutes after its breakthrough from the column. This phenomenon is generally known as the rollover effect and is shown in Figure 10-1. The opposite phenomenon, the displacement of water by organic compounds, is generally observed on activated carbons.

To design an adsorptive air purification system properly, both pure component and mixture adsorption data for the contaminants are required. Although the experimental data for pure components are available for a number of contaminants, due to increased experimental complexities, often the mixture data are not readily available in the literature. Therefore, models or correlations are crucial to the design of multicomponent systems. These models should be capable of estimating the mixture data from the pure component data.

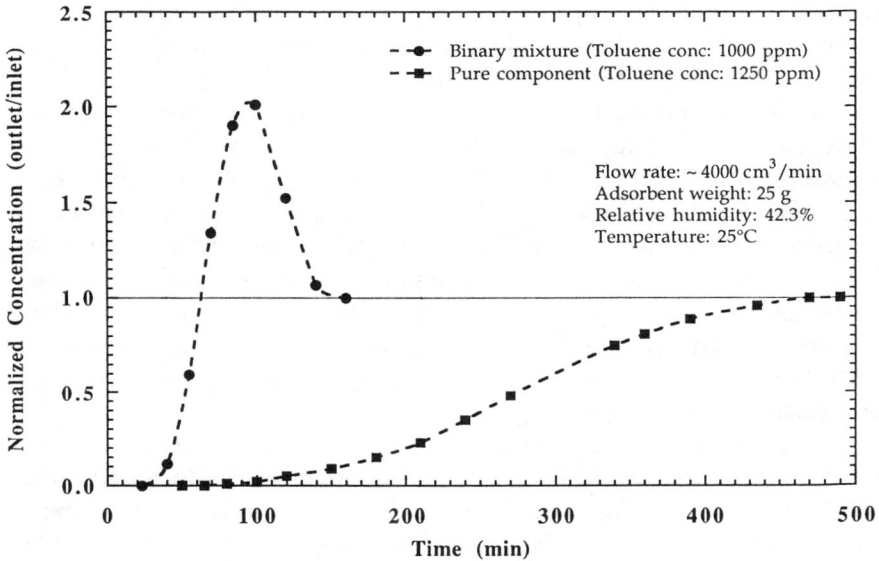

Figure 10-1. Displacement of Toluene by Water Vapor During Adsorption from Their Binary Mixture in a Molecular Sieve 13X Bed.

10.2 CO-ADSORPTION BY SOLID DESICCANTS

Silica gel, activated alumina, and molecular sieves are primarily used as solid desiccants to dehumidify gas streams. Although they have been employed extensively for natural gas dehydration, dehydration of unsaturated hydrocarbons, and adsorption of water, hydrogen sulfide (H_2S) and carbon dioxide (CO_2) from natural gas, their use in indoor air dehumidification systems is more recent.

The use of a regenerable desiccant system both to control humidity and to remove contaminants from the cabin of Space Shuttle was investigated by Amazeen[1]. He evaluated silica gel and molecular sieve 3A, 4A, 5A and 13X to determine their capability to adsorb water vapor and CO_2. Silica gel was found to be the best desiccant, while molecular sieve 4A had the best capacity for CO_2. Amazeen also optimized the bed configurations and assessed the reliability of the system both theoretically and experimentally. Lunde and Kester[2] tested a new sorbent to control water vapor and CO_2 for potential application in the Space Shuttle cabin. Performance tests showed that such material (designated as HS-C) can adsorb both water vapor and CO_2 to provide the desirable concentration level in the cabin. A 2000-hour life test, which included sensitivity to clean solvent vapors, vibration resistance and flammability, was successful.

Gas Research Institute (GRI) of Chicago, Illinois is one of the leaders in developing solid-desiccant-based systems for commercial use, such as in homes, hotels and supermarkets. One of their objectives is to enhance the contaminant-removal capabilities of such systems. A number of projects have been initiated by GRI to study these systems for their capability to remove a variety of contaminants. Relwani and Moschandreas[3] evaluated the performance of a commercially available rotary desiccant unit that employed either silica gel or a molecular sieve by determining its capabilities for removing carbon monoxide (CO), nitrogen dioxide (NO_2), and sulfur dioxide (SO_2), along with water vapor.

Later, Relwani et al.[4] also conducted a similar test with a bench scale apparatus as well as with a rotary desiccant unit that included some unspecified hydrocarbons along with the above contaminants. Experiments were conducted at various inlet contaminant concentrations, inlet humidities of air, and temperature. Factors that affect the regeneration of desiccants also were studied in their experiments. The concentrations of contaminants in an airstream were 0.3 parts per million (ppm) to 0.5 ppm for NO_2, 2 ppm to 3 ppm for CO, 20 parts per billion (ppb) to 50 ppb for SO_2, and 8 ppm to 12 ppm for hydrocarbons. The relative humidity (RH) of the airstream was varied from 10% to 25%. The molecular sieve exhibited higher adsorption capacities for NO_2, CO and SO_2 than silica gel both in the bench and pilot scale experiments. Silica gel demonstrated a higher adsorption capacity for hydrocarbons than the molecular sieve in the bench scale experiments. Silica gel was also easier to regenerate.

The gas-fired desiccant system developed by Novosel et al.[5] with Cargocaire Engineering Corporation is currently being used by 14 supermarket chains throughout the U.S. Furthermore, the system has been specified as one of several alternative systems applicable for military commissaries. Although a systematic study was not conducted to monitor the enhanced comfort in the places where these units were installed, random interviews of the occupants showed a positive response. In a later study, Novosel et al.[6] provided a conceptual workframe of a desiccant-based air conditioner to actively control air quality in residences. From the published report in the literature, the authors expected that the silica-gel-based system could remove formaldehyde and other VOCs during the dehumidification process.

In the last several years, Hines, Ghosh and their co-workers have evaluated a number of desiccant materials for their capability to adsorb various air contaminants. Kuo et al.[7] studied the adsorption of ammonia on Davison grade 59 silica gel by using a packed bed. Breakthrough curves were obtained at 25°C, 40°C and 60°C by passing a helium gas stream containing ammonia through the bed. The equilibrium isotherm data were calculated from breakthrough curves. A significant amount of ammonia can be adsorbed on silica gel under these conditions. Later, they[8-10] studied the adsorption characteristics of various chlorinated hydrocarbons, and aldehydes on Davison grade 40 silica gel in the temperature range of 15°C to 35°C. The adsorption capacity of silica gel for various contaminants at room temperature (25°C) is shown in Table 10-1. As can be seen from Table 10-1, silica gel has a good capacity for a number of contaminants when each one of these is present as a pure component in the vapor phase. However, in a real indoor environment, any removal of contaminants has to be accomplished in the presence of water vapor. Therefore, the uptake of silica gel for these contaminants in the presence of moisture is most important.

Ghosh and Hines[11] studied extensively the contaminant-removal capabilities of silica gel and molecular sieve 13X in the presence of water vapor. Selected contaminants that are representative of a broad spectrum of indoor contaminants were used to provide an assessment of desiccant systems. The representative contaminants were selected on the basis of their concentration indoors, frequency of occurrence, and associated health effects. These contaminants include 1,1,1-trichloroethane (representative of chlorinated hydrocarbons), toluene (representative of aromatic hydrocarbons), CO_2 (representative of combustion products), radon and formaldehyde. CO_2 is not considered to be an indoor contaminant from a health perspective, however, its concentration is the highest among all contaminants.

Although the desiccants have a greater affinity for water vapor, the contaminants investigated in this study were able to compete with water vapor for adsorption sites on the desiccant. The presence of CO_2 in the gas mixture reduced the water uptake capacity of molecular sieve 13X but did not influence the water uptake of silica gel significantly. Although molecular sieve 13X was found to have a greater capacity for CO_2 in the presence of water than silica gel, its capacity in the presence of water vapor was significantly lower than the capacity for the pure CO_2.

Water vapor broke through the silica gel bed earlier than toluene. As

Table 10-1. Adsorption Capacity of Silica Gel for Selected Chlorinated Hydrocarbons and Aldehydes at 25°C.

Pollutants	Concentration in air (ppm)	Adsorption capacity (mg/g of silica gel)
Methyl chloride	1 10 100	0.064 0.607 4.090
Methylene chloride	1 10 100	0.206 1.975 14.630
Chloroform	1 10 100	0.555 5.322 39.635
Carbon tetrachloride	1 10 100	1.030 9.806 68.809
1,1,1-Trichloroethane	1 10 100	1.337 12.709 87.969
Tetrachloroethylene	1 10 100	9.358 77.659 297.930
Acetaldehyde	1 10 100	0.834 7.807 48.066
Propanal (Propionaldehyde)	1 10 100	4.383 33.802 103.911
Butanal (Butyraldehyde)	1 10 100	4.967 39.227 127.494

Source: Hines et al.[18]

the experiment progressed, water eventually displaced the adsorbed toluene from the pores. In a molecular sieve bed, toluene broke through before the water and was also displaced by water vapor as the experiment progressed. As expected, the capacity for adsorbing both water and toluene decreased, but the reduction was greatest for toluene. The decrease in uptake capacity of silica gel for both toluene and water vapor from their binary mixtures is shown in Table 10-2. Similar results were obtained when a molecular sieve was used. The adsorption temperature also affected the uptake of hydrocarbons when the RH of the gas stream was kept constant (refer to Figure 10-2).

Molecular sieve 13X exhibited a catalytic property for 1,1,1-trichloroethane, decomposing it to more toxic vinylidene chloride and hydrogen chloride. The catalytic property was observed when 1,1,1-trichloroethane was present both as a single component and with water vapor in a gas mixture. However, the rate of the reaction decreased in the presence of water vapor. The concentration of vinylidene chloride in the outlet airstream decreased to almost zero after attaining a maximum value (refer to Figure 10-3). Such a catalytic property was not observed on silica gel. Contaminant-removal capabilities of molecular sieve and silica gel for some of these contaminants when the RH of the airstream was 62.5% are shown in Figure 10-4. The adsorption characteristics of other contaminants in the presence of water are discussed in more detail in the report by Hines and Ghosh[11].

Activated aluminas have been in use for more than 25 years in a variety of drying operations for both gaseous and liquid streams, but they have not been employed in desiccant-based air conditioning systems. The initial granular form of activated alumina did not prove satisfactory because of excessive particle breakdown. With the recent introduction of spherical products, activated alumina is finding increasing application as a desiccant.

Table 10-2. Comparison of Adsorption Capacities of Toluene and Water Vapor as Pure Components and from Their Binary Mixtures on Silica Gel.

| Temperature | Concentration (ppm) | Toluene | | | Water Vapor | | | |
		Adsorption Capacity from Binary Mixtures (g/g)	Adsorption Capacity as Pure Component (g/g)[a]	Capacity Reduction (%)	Relative Humidity (%)	Adsorption Capacity from Binary Mixtures (g/g)	Adsorption Capacity as Pure Component (g/g)	Capacity Reduction (%)
15°C (59°F)	1000	0.120	0.212	43.4	22.4	0.0848	0.168	49.5
	1100	0.086	0.220	60.9	39.0	0.178	0.290	38.6
	1000	0.046	0.210	78.1	58.5	0.215	0.332	35.2
25°C (77°F)	1000	0.091	0.178	48.9	16.1	0.059	0.110	46.4
	1000	0.0571	0.174	67.2	42.3	0.215	0.268	19.8
	1000	0.0395	0.173	77.2	63.3	0.258	0.322	19.8
35°C (95°F)	1100	0.0665	0.141	52.8	23.4	0.095	N/A	N/A
	1000	0.0375	0.133	71.8	43.2	0.218	N/A	N/A

[a] Isotherm data of Yeh[64]
N/A- Not Available

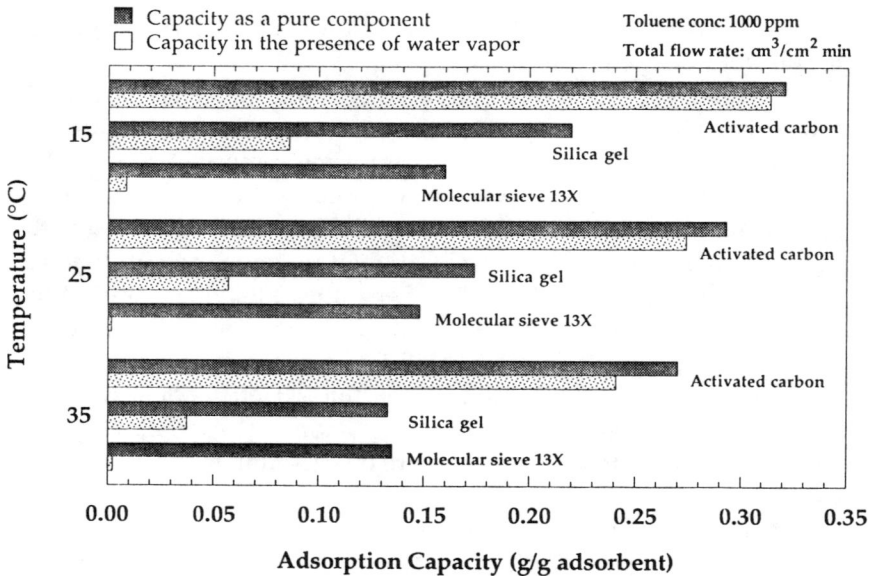

Figure 10-2. Adsorption Capacities of Adsorbents at Different Temperatures for Toluene at a Relative Humidity of 42%.

Activated aluminas are also used for the removal of various contaminants from gases such as trace fluorides, chlorides, H_2S, alcohols and ethers. However, a major application of activated aluminas is in the removal of fluorides from alkylate streams and of H_2S and other sulfur compounds from natural gas by the Claus process. These compounds are removed primarily via chemical oxidation reactions rather than by adsorption. Due to their surface properties, activated aluminas are now being investigated more extensively for use in removing indoor air contaminants.

10.3 CO-ADSORPTION BY ACTIVATED CARBONS

Activated carbon has been used extensively for industrial gas-cleaning operations, but its use for indoor air-cleaning applications is rather recent. Activated carbon has the greatest adsorption capacity for organic molecules because of the non-polar nature of its surface as compared to other solid adsorbents, such as silica gel and molecular sieve. Conse-

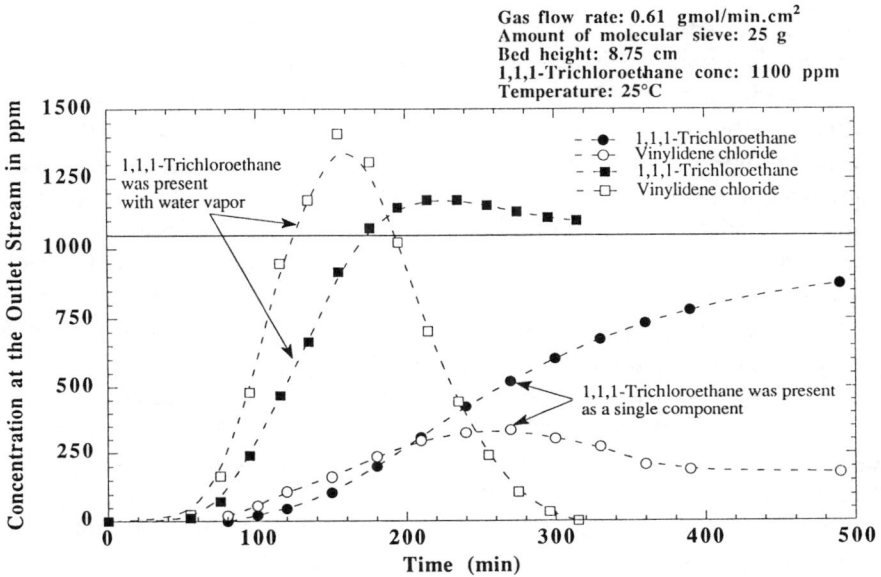

Figure 10-3. Conversion of 1,1,1-Trichloroethane to Vinylidene Chloride Due to Catalytic Effect of Molecular Sieve 13X.

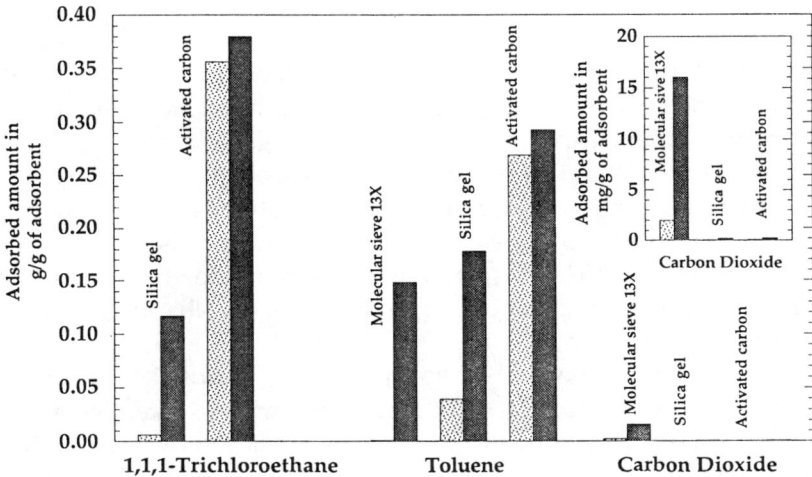

Figure 10-4. Adsorption Capacities of Various Adsorbents for Some Selected Indoor Air Contaminants at 25°C and at a Relative Humidity of 62.5%.

quently, its affinity for water vapor is much lower. This makes activated carbon especially attractive for cleaning indoor air.

In indoor environments, the removal of contaminants has to be accomplished in the presence of moisture. The RH in indoors can normally range from 20% to 80%. However, when designing a carbon-based air cleaner, the assumption has often been made that water vapor does not interfere with the adsorption of organic molecules. This assumption can lead to a significant error in the design and subsequently to poor performance of the bed. When the RH of air is below 40%, activated carbon adsorbs a negligible amount of water. If the RH of air exceeds 40%, activated carbon can adsorb a significant amount of water vapor, which can severely reduce the adsorption capacity for organic compounds. Okazaki et al.[12] obtained adsorption isotherms for water vapor and several organic vapors, including acetone, methanol, benzene, and toluene, on two activated carbons, and found that the presence of water vapor in the airstream decreased the adsorption capacities for organic molecules. Therefore, the designer of a carbon-adsorption system must also be familiar with the water-adsorption characteristics of carbons. Also Hassan et al.[13] pointed out that the amount of water adsorbed can vary from one batch of activated carbon to another.

Activated carbon based adsorption systems have been used effectively to control organic vapors in the workplace. These systems have been designed using experimental data. Hobbs et al.[14]; conducted an extensive survey of the industrial applications of vapor-phase activated carbon adsorption and attempted to develop a database on the basis of information attained from manufacturers of organic compounds. However, they noted that data obtained for industrial applications are often incomplete or are not reported in the open literature because of company restrictions regarding proprietary information. Using these data, attempts have been made to design portable air-cleaners that can be used for cleaning the air in a single room. Several types of portable air-cleaners are already available commercially.

Since concentrations of air contaminants in industrial effluents are much higher than indoor concentrations, care must be taken when interpreting, analyzing, and in the subsequent application of these data to design equipment for indoor environments. Systematic studies that deal with the removal efficiency, humidity effects on the adsorption process, and the competitive adsorption of various contaminants are not reported in the literature. Also, research studies on the use of carbon-filter units for indoor environments such as homes, offices, schools and hospitals have not been reported. Therefore, often the design of indoor carbon-based air cleaning systems is frequently not based on adsorption data, but rather on the myth that carbon can adsorb any organic compound under any circumstance. An excellent review of the literature on the adsorption capacities of activated carbons for organic compounds has been provided by Hines et al[15]. Only some key articles that dealt with binary or multicomponent adsorption are discussed here.

Nelson and his co-workers[16] evaluated the service life of several respirator cartridges that contained activated carbons by measuring the breakthrough point for a number of organic vapors at 22°C. The time required for the outlet concentration to reach 10% of that of the inlet stream, which they called 10% breakthrough time, was reported for various experimental conditions. They also noted that the 10% breakthrough time for a number of organic compounds was attained earlier as the RH of the airstream was increased above 50%. Below 50% RH, the effect was marginal.

The behavior of several organic compounds in their binary mixtures were studied by Reucroft et al.[17] The binary mixtures employed in their study included chloroform-carbon tetrachloride, benzene-n-hexane, chloroform-methylene chloride, and n-hexane-methylene chloride. It was noted that the higher-molecular-weight organic compounds had the greatest adsorption capacity. The adsorption uptake was highest for carbon tetrachloride, followed by chloroform, methylene chloride, n-hexane, and benzene. The total amount adsorbed from the binary mixtures was always between the pure component adsorption capacities of the two chemicals. As an example, for chloroform-carbon tetrachloride mixtures, the amount adsorbed from their mixtures was less than that of carbon tetrachloride alone but was greater than that of chloroform.

Hines and Ghosh[11] also studied the adsorption characteristics of several indoor organic compound contaminants on Calgon BPL activated carbon under both static and dynamic conditions. The equilibrium adsorption capacity of BPL carbon for toluene in the presence of water did not decrease significantly (refer to Table 10-3), but the shape of breakthrough curves changed dramatically when compared with the pure component breakthrough curve. The breakthrough curves for toluene when present as a single component and in the presence of water are shown in Figure 10-5. The rate of adsorption for toluene decreased in the presence of water vapor, resulting in a flattening or tailing of the breakthrough curve. Therefore, care must be taken when designing a carbon adsorber for cleaning indoor air. The design procedure for a carbon bed when removing a single component has been discussed by Hines et al.[18]

Table 10-3. Adsorption Capacities of Toluene and Water Vapor as Pure Components and from Their Binary Mixtures on Activated Carbon.

Temperature	Concentration (ppm)	Toluene			Water Vapor			
		Adsorption Capacity from Binary Mixtures (g/g)	Adsorption Capacity as Pure Component (g/g)[a]	Capacity Reduction (%)	Relative Humidity (%)	Adsorption Capacity from Binary Mixtures (g/g)	Adsorption Capacity as Pure Component (g/g)[b]	Capacity Reduction (%)
15°C (59°F)	1000	0.306	0.316	3.2	24.0	0.0024	0.00492	51.2
	1100	0.314	0.321	2.2	40.1	0.0041	0.0155	73.5
	1000	0.300	0.316	5.1	60.2	0.00821	0.118	93.0
25°C (77°F)	1000	0.284	0.293	3.1	16.8	0.0015	0.0053	71.7
	1000	0.274	0.293	6.5	42.3	0.0029	0.023	87.4
	1000	0.269	0.293	8.2	62.9	0.0051	0.24	97.9
35°C (95°F)	1000	0.255	0.270	5.6	24.4	0.0009	0.0057	84.2
	1000	0.241	0.270	10.7	44.1	0.0019	0.0155	87.7

[a] Isotherm data of Yeh[64]
[b] Isotherm data of Hassan[65]

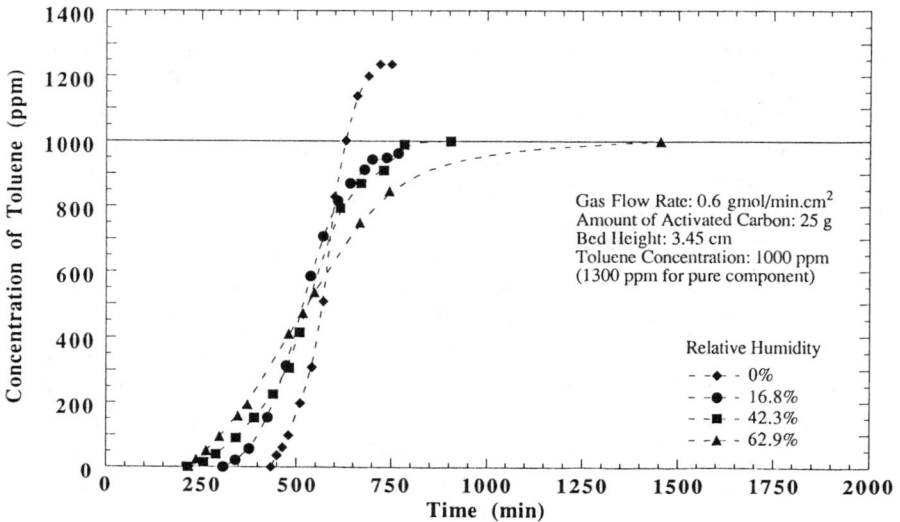

Figure 10-5. Comparison of Breakthrough Characteristics of Toluene on Activated Carbon at 25°C and at Different Relative Humidities.

10.4 FUNDAMENTALS OF ADSORPTION

Adsorption involves the transfer of a material from one phase to a surface, mainly from a gas or liquid to a solid surface, where it is bound by intermolecular forces. The design of an effective adsorption process, such as a desiccant-based dehumidification system, requires information about the amount of gases or vapors adsorbed as a function of their concentration in the gas-phase at different temperatures, which is generally referred to as an adsorption isotherm or equilibrium adsorption data. When the amount of gas or vapor adsorbed on a solid surface is plotted corresponding to its gas-phase concentration at a particular temperature, curves representing the isotherm data can be obtained. The great majority of the isotherms observed to date can be classified into the five types as shown in Figure 10-6. The different shapes are an indication of the actual adsorption mechanism, which is related to the physical and chemical properties of the adsorbate and to the surface characteristics of the adsorbent.

Adsorption isotherm data can be obtained using either a static or dynamic system. In a static system, which is most frequently used to obtain the pure component isotherm data because of its simplicity and rapidity, the amount of gas or vapor adsorbed on the solid surface is measured as a function of pressure or concentration, when the solid adsorbent is exposed to an atmosphere containing the adsorbate at constant temperature. In a flow system, a stream containing the adsorbate (contaminant) is passed through a fixed-bed of solid adsorbent. As the bed becomes saturated, the adsorbate concentration in the outlet stream becomes equal to that of the inlet stream.

The data obtained from such systems are expressed in terms of the concentration profile of the adsorbate in the effluent gas stream as a function of time, and is typically referred to as a "breakthrough curve." The amount of a contaminant adsorbed in the bed can be obtained either by weighing the bed after the experiment or from the area behind the breakthrough curve. This provides an equilibrium point on the isotherm curve. Therefore, a series of runs have to be conducted to obtain a complete isotherm curve.

10.5 ISOTHERM MODELS

The selection of a model or correlation to accurately represent the adsorption data will depend on whether the adsorption is physical or

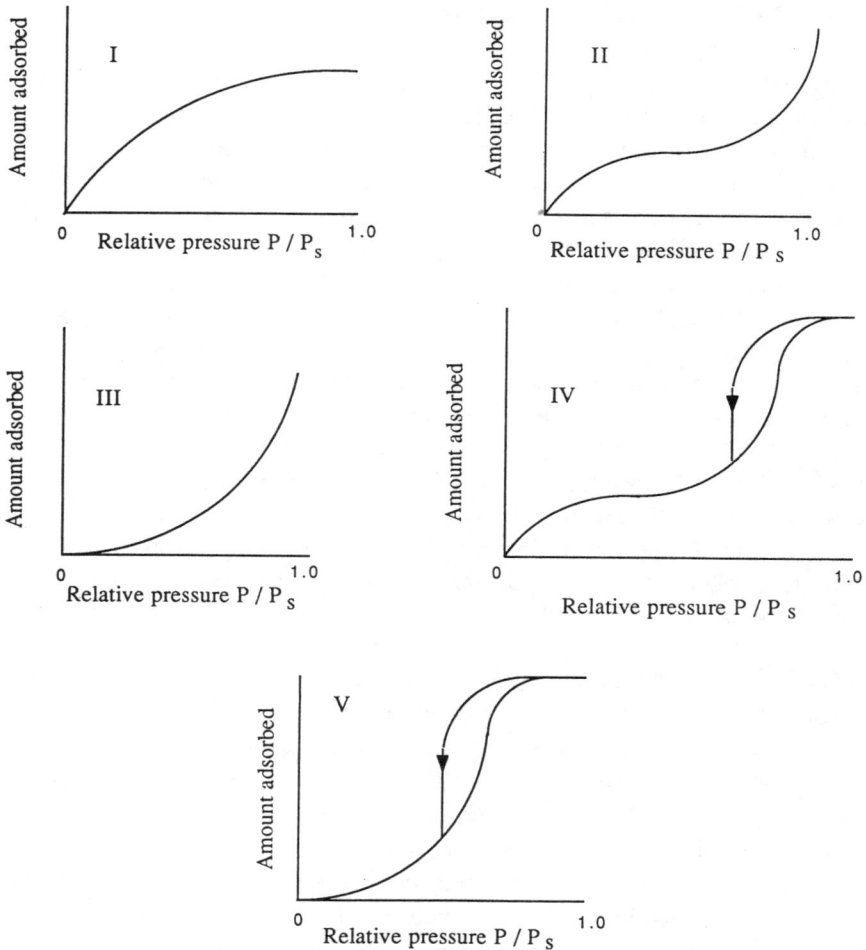

Figure 10-6. Brunauer's Classification of Adsorption Isotherms (P: System Pressure, and P_S: Saturation Pressure of Adsorbate at System Temperature).

chemical in nature. Physical adsorption results when the adsorbate adheres to the surface by the van der Waals' forces (i.e., by dispersion and Coulomb forces). During adsorption, a quantity of heat (heat of adsorption) is released. The quantity of heat released during physical adsorption is approximately equal to the heat of condensation, resulting in adsorption being frequently described as a condensation process. The primary difference between physical adsorption and chemisorption is the nature of the bond that is formed between the adsorbed molecule and the adsorbent surface. Chemisorption is characterized by a sharing of electrons between the adsorbent and adsorbate, which results in the liberation of a quantity of heat that is approximately equal to the heat of reaction. Although chemical and physical adsorptions are characterized by different thermal effects, a clear distinction between the two adsorption mechanisms does not exist. The adsorption of water vapor and other indoor contaminants on both desiccants and activated carbons is generally physical in nature, as indicated by the heat of adsorption.

The equilibrium isotherms shown in Figure 10-6 for the adsorption of vapors were classified into five principle types by Brunauer[19]. Type I is characterized by a monotonic increase to a limiting adsorption capacity that corresponds to the formation of a complete monolayer, and is generally known as the Langmuir type. The adsorption isotherms of water vapor and contaminants on silica gel are generally of Type I. This type is also found for systems in which the adsorbate is chemisorbed. A type II isotherm is characteristic of the formation of multiple layers of adsorbate molecules on the solid surface. This type, which is also known as the BET after Brunauer, Emmett and Teller[20], has been found to exist mainly for nonporous solids. However, the data of water and other chemical vapors on molecular sieves usually exhibit Type II shape. Type III isotherms, although similar to Type II because they have been observed for nonporous solids, are relatively rare. The shape of Type III isotherms also suggests the formation of multilayers. Type IV and V isotherms are considered to reflect capillary condensation since they level off when the saturation pressure of the adsorbate vapor is reached.

10.5.1 Single Component Monolayer Models

A number of mathematical models have been proposed to describe the adsorption process. However, a single model is not available that can describe all five types of isotherms. Most models have been developed for Type I and II isotherms, since a great majority of isotherms are one of these types. In addition to monolayer and multilayer adsorption equations, models have been developed to describe situations in which the adsorption occurs either locally on specific sites or is mobile over the surface of the adsorbent. Consideration also has been given to cases in which the adsorbed molecules interact both with the surface and with each other. Detailed discussions of the various isotherms and the mechanisms of adsorption have been provided by several researchers.[21-24]

One of the most frequently used equations for a Type I isotherm was developed by Langmuir[25]. This equation was developed assuming no interaction between molecules on the surface and that the surface is homogenous. Although these are not valid assumptions for most of the adsorbents, it provides a reasonable description of Type I systems and its use is often justified based on its ability to fit equilibrium data. The Langmuir equation can be expressed as:

$$q_A = \frac{QKC_A}{1 + KC_A}$$

(10-1)

where

q_A: equilibrium uptake of adsorbate A by the adsorbent corresponding to concentration C_A,

Q: weight of adsorbate contained in the monolayer on the surface,

C: concentration of adsorbate A in the fluid phase in equilibrium with the concentration q_A on the solid, and

K: constant.

Another frequently used isotherm is the Freundlich equation given as:

$$q_A = K(C_A)^{1/n}$$

(10-2)

where

q_A: equilibrium uptake of adsorbate A by the adsorbent corresponding to concentration C_A,

C_A: concentration of A in the surrounding fluid that is in equilibrium with the concentration q_A on the solid,

K: constant, and

n: constant.

The Freundlich equation was derived by Zel'dovich[26] from a consideration of the decrease in the number of adsorption sites. Even though it is not based on a rigorous theoretical background, the Freundlich isotherm gives an adequate description of adsorption equilibrium of many air contaminants.

It is important to note that the Langmuir equation reduces to a linear form (frequently identified as the Henry's law equation) as the fluid phase concentration approaches zero, whereas the Freundlich equation does not. The Henry's law equation is given as:

$$q_A = K_h C_A \tag{10-3}$$

where K_h is the Henry's law constant. Very low pressure data are required to obtain K_h. When fitting equilibrium adsorption data for indoor air contaminants, which are usually present at low concentrations, one should be cautious about using the Freundlich equation.

Jovanovic[27] proposed another simple equation that applies to Type I isotherms, which can be expressed as:

$$V = V_m \left[1 - \exp\left(-a\frac{P}{P_s} \right) \right] \tag{10-4}$$

where V_m is the volume adsorbed in the monolayer, P_s is the saturation pressure of the adsorbate at the system temperature, and "a" is a constant that describes adsorption in the monolayer and can be derived from the kinetic theory of gases. However, in most applications, both the constants V_m and "a" are obtained by nonlinear regression of the experimental data. Although not as widely used as the isotherms discussed earlier, the Jovanovic equation has been used extensively in the development of several more complicated and accurate isotherm models that apply to heterogeneous surfaces.

Each of the above models applies to Type I isotherms, but frequently fails to fit the equilibrium data with suitable accuracy, particularly in the low-pressure region. This is often attributed to the heterogeneous nature of the adsorbent surface. As a result, recently a number models have been proposed for Type I systems that take into consideration the energy of an adsorption site. It is well known that for a large class of solid-vapor systems, the surface heterogeneity plays an important role in determining adsorption characteristics, and its effects require adequate treatment. The general approach used to describe adsorbent heterogeneity is to postulate that the heterogeneous surface exhibits a distribution of adsorptive potentials which are either grouped in patches or distributed randomly on the surface. Even small variations in the adsorption potential have been shown to influence the adsorption behavior.

According to Ross and Olivier[22], the overall isotherm can be obtained by integrating the contribution of each patch over the energy distribution range. The adsorption isotherm is thus given as:

$$Q(P,T) = \int_0^\infty Q_1(P,T,\varepsilon) E(\varepsilon) d\varepsilon$$

(10-5)

where $E(\varepsilon)$ is a probability distribution function for the energy patch on the surface and $Q(P,T)$ is the overall adsorption isotherm on the heterogeneous adsorbent. P and T are the system temperature and pressure, respectively. The term $Q_1(P,T,\varepsilon)$ describes the local adsorption isotherm for homotattic sites of adsorptive energy, ε. Both Q and Q_1 are amounts adsorbed per unit mass of adsorbent. Several studies have been carried out to develop analytical functions for $Q(P,T)$ since they are useful for data extrapolation and for the modeling of adsorption processes. Sircar[28] developed an expression for $Q(P,T)$ by assuming that the Langmuir model could be used to represent the local isotherm and that the gamma probability density function described the energy distribution of the adsorbent surface. Sircar's model successfully described the adsorption of various gases on activated carbons and zeolites over wide ranges of pressure and temperature. The resulting equation, however, is somewhat cumbersome to use. Sircar[29] later developed a simpler model that had the same degree of versatility by using the Jovanovic local isotherm model in conjunction with the gamma distribution function.

Hines et al.[9] developed an adsorption equation to describe the adsorption of gases on energetically heterogeneous surfaces, using the Jovanovic model as the local isotherm, in conjunction with a modified Morse-type distribution function to describe the energy distribution of the heterogeneous surface. The Hines et al. equation is given as:

$$Q(P,T) = Q_m \left[1 - \frac{K_1 K_3}{K_3 - K_1 K_2} \left(\frac{1}{P + K_1} - \frac{K_2}{P + K_3} \right) \right]$$

(10-6)

where Q_m is the saturation adsorption capacity and the K-values are constants. The adjustable model parameters (Q_m, K_1, K_2, K_3) were determined for each data set by using a nonlinear regression analysis. Their model successfully represents experimental adsorption isotherm data for several gases and vapors on different heterogeneous adsorbents over a wide range of temperatures and pressures. Furthermore, it is easy to use, and it provides a useful tool for accurately correlating isotherm data on heterogeneous surfaces. Also, the model parameters can be expressed as a function of temperature, allowing the extrapolation of the data over a moderate temperature range.

10.5.2 Single Component Multilayer Models

Although Type II isotherms, which occur due mainly to the formation of multiple adsorbed layers, have been observed for a large number of adsorbate-adsorbent systems, only a limited number of models are available to correlate the data. These models often fail to fit equilibrium adsorption data over a wide range of pressures.

The first multilayer model was developed by Brunauer et al.[20] considering the same basic assumptions of the Langmuir equation, and that when multilayers are formed, the heat of adsorption for these layers are different from the first layer. Their model is best known as the BET equation and is given as:

$$\frac{P}{V(P_s - P)} = \frac{1}{V_m C} + \frac{(C-1)P}{C V_m P_s}$$

(10-7)

where V is the volume adsorbed at pressure P, V_m is the volume occupied in a monolayer, P_s is the saturation pressure at the system temperature, and C is a constant. The BET equation is usually valid over a relative pressure range, P/P_s, from 0.05 to 0.35. Therefore, it is generally not used to fit adsorption data, but it is widely used as a method for determining the surface area of adsorbents. A procedure for estimating surface area using the BET equation can be obtained in Hines and Maddox[30]. Ghosh and Hines[10] compared the correlation capability of the Langmuir, BET, Freundlich, and Hines et al. models using the data for aldehyde-silica gel systems. The average absolute percent error, maximum positive error, and maximum negative error for these systems are shown in Table 10-4. The Hines et al and the Freundlich models provided a better correlation of the data than the Langmuir and BET models.

Table 10-4. Comparison of Model Correlations.

System	Temperature (°C)	Absolute Average Error (%)				Maximum Positive Error (%)[a]				Maximum Negative Error (%,)[a]			
		Hines et al.	Langmuir	BET	Freundlich	Hines et al.	Langmuir	BET	Freundlich	Hines et al.	Langmuir	BET	Freundlich
Acet-	14.0	0.58	12.49	15.40	1.71	2.06	72.69	8.78	7.98	1.17	9.25	105.95	2.37
aldehyde-	25.2	0.40	8.51	5.62	0.93	0.87	65.04	54.62	2.58	2.49	6.23	4.07	1.53
Silica gel	33.5	0.46	9.82	6.50	0.89	1.01	59.69	50.67	3.23	1.20	9.00	5.44	1.34
Propion-	9.0	1.73	5.15	142.36	1.27	2.10	42.92	1753.71	1.72	4.30	3.08	190.95	3.37
aldehyde-	24.0	0.26	6.61	8.23	0.88	0.62	52.01	3.83	5.66	0.82	5.64	54.33	1.54
Silica gel	31.3	0.87	6.54	1.96	1.00	1.15	39.88	1.71	4.42	1.46	7.01	5.73	1.53
Butyr-	15.2	1.55	6.07	94.38	1.19	2.81	49.02	743.21	1.69	2.77	4.63	222.09	3.74
aldehyde-	26.3	0.66	3.44	10.01	0.61	0.88	23.68	5.64	1.99	1.78	2.57	70.58	1.28
Silica gel	35.2	0.27	5.12	5.57	0.82	0.48	36.92	4.31	3.05	0.85	3.37	33.69	1.52

[a]Error = [(experimental − calculated)/(experimental)] × 100.

Source: Ghosh and Hines[10]

Jovanovic[27] also proposed an isotherm for multilayer adsorption for correlating data over a relative pressure range of $0.2 < P/P_s < 0.7$. This model is given as:

$$V = V_m \left[1 - \exp\left(-a\,\frac{P}{P_s} \right) \right] \exp\left(b\,\frac{P}{P_s} \right)$$

(10-8)

where "a" was defined earlier for the Jovanovic monolayer model and "b" is a constant and is related to uptake in the second and higher layers.

Ghosh[31] developed a multilayer adsorption model for heterogeneous surfaces by using the same method described earlier for the development of the model of Hines et al., except that the Jovanovic multilayer adsorption isotherm equation was employed as the local isotherm. Ghosh's equation is written as:

$$Q(T,P) = Q_m \left[1 - \frac{K_1 K_3}{K_3 - K_1 K_2} \left(\frac{1}{K_1 + P} - \frac{K_2}{K_3 + P} \right) \right] \frac{\exp(-b_0 P)}{\varepsilon_m P} \left[\exp\left(\varepsilon_m P - 1 \right) \right]$$

(10-9)

where K_1, K_2 and K_3 are energy parameters in a Morse-type probability distribution function, Q_m is the adsorption capacity at saturation, b_0 is the limiting value of the constant b given in the Jovanovic equation, and ε_m is the energy parameter for second and higher layers in the adsorbed phase. Ghosh's model was tested for several systems. The adsorbents for the systems tested were microporous, and the data for these systems could not be accurately described with the existing homogeneous isotherm equations. Eqn. (10-9) provided excellent correlation of the data for all systems tested. This model is recommended for Type II isotherms.

10.5.3 Vacancy Solution Model

Danner and co-workers[32-34] developed several isotherm equations based on the vacancy solution theory. The effect of surface heterogeneity and the interaction between adsorbate-adsorbent and adsorbate-adsorbate were taken into consideration by introducing activity coefficients in the adsorbed phase. In this approach, the solid surface is considered to be composed of a vacancy (species n) and an adsorbed species (species a), and the bulk vacancy solution is very dilute. However, the vacancy is an imaginary entity that is defined as the vacuum space which acts as the solvent for the system. By equating the chemical potential of the adsorbed and gas phases, an equation for the adsorbed phase is obtained. The chemical potential of the adsorbed phase contains an activity coefficient which accounts for the nonideality in the adsorbed phase. Various activity coefficient models have been used, however, in the latest version of the theory the Flory-Huggins activity coefficient equation is employed. This isotherm equation is given as:

$$P = \left(\frac{n_a^\infty}{b_a} \frac{\theta}{1-\theta} \right) \exp \left(\frac{\alpha_{av}^2}{1 + \alpha_{av} \theta} \right)$$

$$(10\text{-}10)$$

where

P: equilibrium adsorption pressure,

n_a^∞:   limiting amount adsorbed of pure component,

b_a: Henry's law constant,

θ: fraction of limiting adsorption, and

α_{av}:   parameter describing nonideality in adsorbed phase induced by interaction between species a and v.

Eqn. (10-10) has three adjustable parameters, n_a^∞, b_a, and α_{av} that are to be determined from the experimental data. If the adsorbed phase is assumed to be ideal ($\alpha_{av} = 0$), the equation reduces to the form of the familiar Langmuir equation. Three parameters can be expressed as a function of temperature, but this increases the number of adjustable parameters to five. The Henry's law constant, b_a, is related to the heat of adsorption at infinite dilution q_a by the following expression:

$$b_a = b_{oa} \exp \left(- q_a / RT \right)$$

$$(10\text{-}11)$$

where b_{oa} is characteristic of the adsorbate-adsorbent system and may be assumed to be independent of temperature. An empirical relationship was used to describe the temperature dependency of n_a^∞.

$$n_a^\infty = n_{oa}^\infty \exp \left(r_a / T \right)$$

$$(10\text{-}12)$$

The temperature dependence of α_{av} was also obtained empirically and is given as:

$$\alpha_{av} = m_a n_a^{\infty} - 1$$

$$(10\text{-}13)$$

Therefore, the five adjustable parameters are b_{oa}, q_a, n_a^{∞}, r_a, and m_a. Experimental data, using at least three different temperatures, should be used to obtain these parameters.

10.5.4. Multicomponent Models

When designing an adsorption system to remove two or more contaminants from a stream, multicomponent adsorption equilibrium data are required. As mentioned earlier, multicomponent data are often not readily available and are also difficult to obtain experimentally. Therefore, multicomponent mixture data need to be estimated from pure component isotherm data by means of a suitable model. A number of such models have been developed, but an exhaustive discussion of all of the existing multicomponent models is beyond the scope of this chapter. The interested reader is referred to the works of Ruthven[23] and Yang[24]. Our discussion will be limited to only selected models.

Markham and Benton[35] extended the single component Langmuir equation to describe adsorption in a multicomponent system. The amount of component i adsorbed from a mixture of n components is given by:

$$q_i = \frac{Q_i K_i P_i}{\left(1 + \sum_{j=1}^{n} K_j P_j\right)}$$

$$(10\text{-}14)$$

where P_i is the partial pressure of component i in the gas phase, Q_i is the amount adsorbed in the monolayer for component i, and K_i is a constant for component i. The values of both Q_i and K_i are obtained from the pure component data of component i. The total quantity adsorbed for all of the components is thus given as:

$$q_T = \sum_{i=1}^{n} q_i = \frac{\sum_{i=1}^{n} Q_i K_i P_i}{\left(1 + \sum_{j=1}^{n} K_j P_j\right)}$$

$$(10\text{-}15)$$

Broughton[36] and Kemball et al.[37] showed that the above equation was thermodynamically consistent if the amounts of each component adsorbed in the monolayer are equal (i.e., $Q_1 = Q_2 = ... = Q_n$). However, for most of the system this is not true. Innes and Rowley[38] suggested that an average value be used for the monolayer coverage. For a binary system the average value is obtained from the expression:

$$\frac{1}{Q_a} = \frac{x_1}{Q_1} + \frac{x_2}{Q_2}$$

(10-16)

where Q_a is the average amount adsorbed in the monolayer, and x_1 and x_2 are mole fractions of the two components on the solid surface. Good agreements with experimental data are obtained by using the averaged value for mixtures with widely different (e.g. over twofold) monolayer adsorbed amounts. Although the above modification provides significant improvement in terms of its ability to fit the mixture data, the model is limited to pure component data that can be fit with the Langmuir equation. Also, for ternary or higher order mixtures, a trial and error procedure will be necessary since the adsorbed phase mole fractions are generally unknown.

Another model that is frequently used in calculation of adsorber dynamics for mixture is called the Loading Ratio Correlation (LRC). This model was developed by combining the Langmuir and Freundlich equations and, therefore, does not have any theoretical foundation. The LRC is given as:

$$\frac{q_i}{Q_i^*} = \frac{K_i P_i^{1/n_i}}{\left[1 + \sum_{j=1}^{n} \left(K_j P_j \right)^{1/n_j} \right]}$$

(10-17)

where q_i is the amount of the component i adsorbed from the mixture, and Q_i^* was defined as the maximum obtainable loading instead of the monolayer capacity. The ratio q_i/Q_i^* for a pure component represents the loading ratio for the adsorbent. It does, however, provide a good fit to experimental data as demonstrated by several researchers[39-41] and, consequently, is used widely in adsorber design. The parameters in the LRC can be obtained from pure component isotherm data. In principal, it is thus possible to predict multicomponent equilibrium.

A number of multicomponent adsorption models have been developed based on the Ideal Adsorbed Solution (IAS) theory[42-45]. The basic assumption of the IAS theory is that the adsorbed phase formed an ideal solution on the solid surface and equilibrium exists between the gas phase and the adsorbed phase for each component of the mixture. Sircar and Myers[46] showed that the models developed by the previous researchers differed from each other primarily in their choice of standard states. At equilibrium, the relationship between the adsorbed and gas phases is given by:

$$Py_i = P_i^\circ x_i \tag{10-18}$$

where P_i° is a function of the spreading pressure (π) and is interpreted as the equilibrium pressure that pure component i should have in the gas phase to produce the same spreading pressure as that of the mixture (π_m) at the same temperature, when adsorbed on the solid surface, and which can be written as

$$\pi_1 = \pi_2 = ... = \pi_m$$

or,

$$\frac{\pi_1 A}{RT} = \frac{\pi_2 A}{RT} = ... = \frac{\pi_m A}{RT} \tag{10-19}$$

where A is the surface area of the adsorbent, and R is the universal gas constant. The term P is the total pressure and y_i and x_i are mole fractions in the gas phase and on the solid surface, respectively. When an N-component gas mixture is in equilibrium with a particular adsorbent, the equilibrium state is defined by the concentrations in the adsorbed phase (n_i, i = 1,2,...N) and the concentration in the gas-phase ($P_i = Py_i$, i = 1,2,...,N). Therefore, the adsorption equilibrium between a gas mixture of N components and an adsorbent can be defined by Eqn. (10-18) and the following system of equations:

$$\frac{\pi_i A}{RT} = \int_0^{P_i^0} \left(\frac{n_i^0}{P} dP \right) \quad i = 1,2,...,N \tag{10-20}$$

$$n_t = \left[\sum_{i=1}^{N} \frac{x_i}{n_i^0} \right]^{-1}$$

$\hspace{9cm}$ (10-21)

$$n_i = n_t \, x_i \qquad i = 1, 2, ..., N$$

$\hspace{9cm}$ (10-22)

n_i is the amount of the ith component adsorbed per unit mass of adsorbent and is given by an isotherm equation.

O'Brien and Myers[47] developed an algorithm based on the IAS theory to predict the adsorbed phase mole fractions from pure component isotherm data. The algorithm required an isotherm equation to represent the pure component data accurately, which at the same time, could be integrated analytically to obtain the spreading pressure. Otherwise, their model required repeated numerical integrations that increased the computation time. Later, Moon and Tien[48] tried to improve the algorithm by reformulating the equations of O'Brien and Myers. Substantial improvement in computing time was reported for mixtures having more than five components. Moon and Tien compared the predictive capability of their model with the algorithm of O'Brien and Myers.

Many researchers[49-52] have pointed out that only a few systems behave ideally in the adsorbed phase. However, the nonideality of the adsorbed phase was taken into account by introducing the activity coefficient in Eqn. (10-18). Assuming that the gas-phase is ideal, Eqn. (10-18) becomes:

$$P y_i = \gamma_i P_i^\circ x_i$$

$\hspace{9cm}$ (10-23)

where γ_i is the activity coefficient. Several methods have been proposed to calculate activity coefficients in the adsorbed phase. One of these is to calculate the activity coefficient for each component from the experimental data of binary mixtures, and then fit the values to various activity coefficient models, such as Wilson, UNIQUAC, and Hildebrand. One of these models can be next used to predict ternary or higher order mixture adsorption equilibria.

Costa et al.[49] used the Wilson and the UNIQUAC type activity coefficient models to estimate the activity coefficient of each component in the mixture, while the binary interaction parameters for these models were obtained from the binary mixture data. Ghosh et al.[52] proposed a method to estimate the binary mixture equilibria by using only pure component data, while concurrently taking into account the nonideality in the adsorbed phase. Although they tested their model for binary mixtures only, it can be extended to multicomponent mixtures.

Cochran et al.[34] also extended their pure component isotherm equation, (Eqn. 10-10) to multicomponent mixtures. Their model is capable of predicting gas mixture equilibria using only the parameters obtained from the pure gas data. For the detailed calculation procedure, interested readers should refer to the book by Yang[24]. A review of other methods to predict multicomponent adsorption has been provided by Sircar and Myers[46]. Jaroniec[53] reviewed the kinetics and the equilibrium state of adsorption for multicomponent systems. Very few experimental data for mixtures containing more than two components are available in the literature. Therefore, all these multicomponent models were not tested rigorously with the mixture data of various types of adsorbate-adsorbent systems. However, it was noted that the model prediction becomes worse as the number of components in the gas mixture increases.

The model prediction for binary mixtures is generally satisfactory, but a deviation of more than 50% has been observed for ternary or higher order gas mixtures. Therefore, extreme care must be taken when choosing a model for calculating multicomponent adsorption equilibria. Detailed comparisons between various multicomponent models are provided by Yang[24] and Ruthven[23]. However, to design adsorbers for multicomponent systems and cyclic gas separation processes, generally extended Langmuir isotherm or the LRC model is used because of the noniterative procedure for the calculation.

10.6 DYNAMIC ADSORPTION

Adsorptive gas separation or purification processes are generally operated under dynamic conditions. A gas stream containing contaminants is flown through an adsorber for a specific period of time after which it is regenerated. Although the operation of each bed is batchwise, the system as a whole is continuous since it is operated in a cyclic steady state. The analysis of adsorption in packed beds is based on the study of effluent concentration profiles, which are usually referred to as breakthrough curves. The shape of concentration profiles are a function of adsorber geometry and operating conditions, and equilibrium adsorption data. These breakthrough curves are obtained by flowing a fluid that contains an adsorbable solute with an initial concentration, C_o, through a packed bed that contains a clean or regenerated adsorbent. As the flow of the fluid continues, the bed becomes saturated with the adsorbate at a given position, and a concentration distribution is established within the bed as shown in Figure 10-7.

Although the solute first appears in the effluent stream at time t_i, the time t_b is most important in the adsorber design. At time t_b, the concentration of the contaminant, C_b, attains the maximum value allowable in the effluent stream. Time t_e is the time at which the bed becomes totally saturated with the adsorbate. At this time the bed is exhausted. However, the bed is generally regenerated after the breakpoint is reached (i.e., after time t_b). It can be readily seen that the area behind the breakthrough curve once the bed become saturated (i.e., after time t_e), represents the quantity

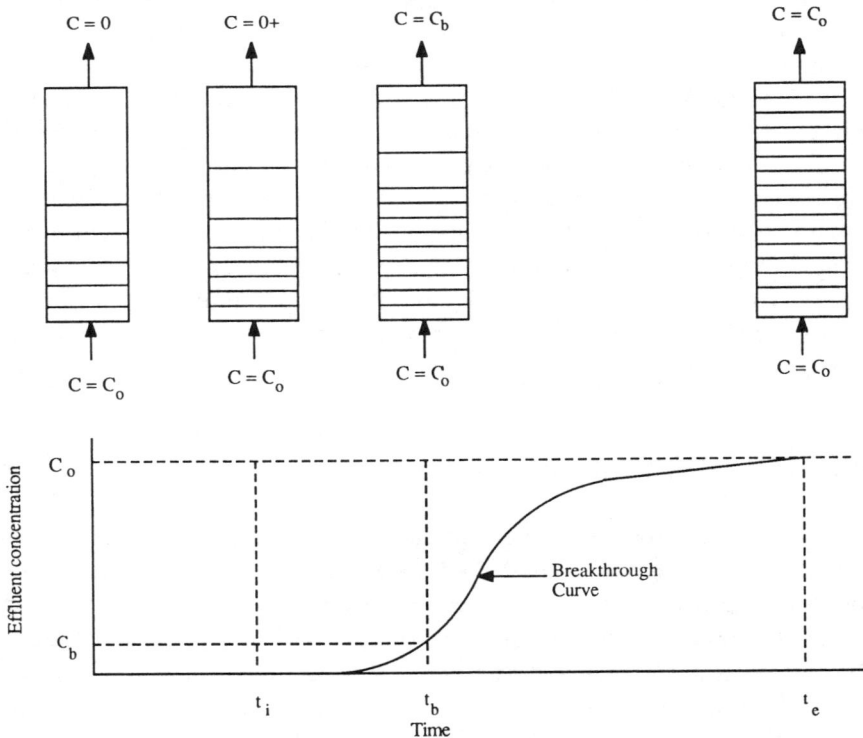

Figure 10-7. The Concentration Profile of an Adsorbate from a Fixed Bed.

of adsorbate retained by the adsorbent contained in the column. This corresponds to a point on the equilibrium isotherm.

As can be seen in Figure 10-7, when the breakpoint concentration C_b is reached, a portion of the bed still remained unused. The effective use of the bed depends on the shape of the isotherms. For the purpose of discussing dynamic adsorption, we will classify the equilibrium isotherms as either (a) favorable, (b) linear, or (c) unfavorable. These are shown in Figure 10-8. If the isotherm is concave in the direction of the fluid phase concentration, as shown by curve (a), layers of high concentration in the bed move faster than layers of lower concentration. This results in the adsorption zone becoming thinner as the wavefront moves through the bed, and gives a breakthrough curve that is self-sharpening. For a favorable isotherm, the breakthrough curve develops and moves through the packed column in a constant pattern. The unfavorable isotherm results in a breakthrough curve that becomes more diffuse as it traverses the bed length, and a significant portion of the bed remains unused. For nonequilibrium adsorption, a favorable isotherm will yield a constant pattern breakthrough curve after a period of time. In most industrial adsorption systems, however, equilibrium between the adsorbate and adsorbent is not reached.

The velocity of the gas stream flowing through the bed also changes the shape of the breakthrough curve and influences the rate of adsorption. At low flow rates, equilibrium conditions are approached, but axial dis-

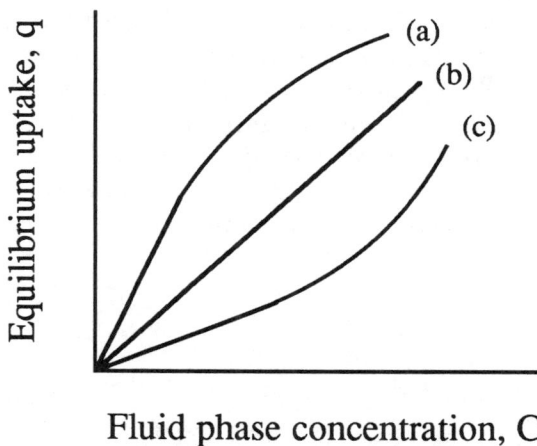

Figure 10-8. Shape of Equilibrium Isotherms.

persion of the adsorbate can be significant and must be considered in the bed design. At lower velocities, the rate of adsorption may be limited by the rate of mass transfer from the fluid phase to the solid surface. For higher flow rates, axial dispersion is typically insignificant, relative to its effect on the rate of mass transfer and its influence on the shape of the breakthrough curve, but adsorbate-adsorbent equilibrium is not attained.

The design of adsorption equipment and the prediction of its performance for a range of operating conditions is still essentially an art based on prior experience. Also, a large amount of experimental data is required to verify the models. Two approaches can be used to solve the mass transfer equation for a packed bed adsorber. In the first approach, a differential equation that describes the transfer of the adsorbate into the column is coupled with the equation that describes the diffusion process into the adsorbent particles. Generally numerical techniques are used to obtain a solution. A detailed discussion of various numerical methods has been provided by Ruthven[23]. In the second approach, the rate of mass transfer of the adsorbate from the fluid to the adsorbent is considered. The overall rate of adsorption is assumed to be controlled by one or a combination of the following mechanisms: (a) external mass transfer to the surface, (b) internal mass transfer through the fluid that fills the pores of the adsorbent, (c) internal mass transfer across the solid surface, and (d) the actual rate of adsorbate uptake. Discussions of the various mechanisms that affect the rate of adsorption in a bed are given by Hines and Maddox[30] and Yang[24].

In indoor environments, contaminants are present at a very low concentration. In modeling an adsorber, a plug flow system in which a trace of an adsorbable species is adsorbed (therefore, the velocity through the bed may be assumed to be constant) from an inert carrier gas (air, in the case of an indoor environment) may be assumed. By considering no radial variation in concentration and neglecting axial diffusion, the mass balance equation for a component i in a mixture of N components can be written as:

$$v\frac{\partial C_i}{\partial z} + \frac{\partial C_i}{\partial t} + \left(\frac{1-\varepsilon}{\varepsilon}\right)\frac{\partial Q_i}{\partial t} = 0 \qquad i = 1,2,...,N$$

$$(10\text{-}24)$$

where v is the fluid velocity through the bed calculated based on empty cross section of the bed, z is the length of the bed, C_i is the concentration of the component i in the bulk fluid, ε is the bed void fraction, and Q_i is the concentration in the solid phase. If instantaneous equilibrium is assumed at all points in the column, the relationship between Q_i and C_i may be expressed by an isotherm equation.

$$Q_i = f(C_i) \tag{10-25}$$

However, as mentioned earlier, in most applications, equilibrium between adsorbate and adsorbent is not reached since a high flow rate is used to reduce the effect of axial dispersion. In this case, the mass transfer resistance is approximated by using a linear driving force rate and is given as:

$$\frac{\partial Q_i}{\partial t} + \frac{K_f a}{\rho_b} + \left(C_i - C_i^*\right) \tag{10-26}$$

where

$K_f a$:   fluid phase mass transfer coefficient,
ρ_b: density of the adsorption bed, and
C_i^*:   concentration of component i in the fluid phase that is in equilibrium with the solid.

Other types of mass transfer mechanisms can be assumed. For a single component, that is when only one adsorbable species is present in the gas stream, analytical solutions are available for different controlling mechanisms in the adsorption process.

Hougen and Marshall[54] solved the mass balance equation assuming isothermal conditions, a linear isotherm, and mass transfer controlled by diffusion through the boundary layer. Rosen[55,56] considered a diffusion resistance to mass transfer in the adsorbent granules along with the same assumption as that of Hougen and Marshall. Analytical solutions are available both in graphical and tabular form. Hiester and Vermeulen[57] solved the equations for a Langmuir isotherm and presented the results graphically and in the form of tables. Van Deemter et al.[58] and Acrivos[59] took into account the effect of longitudinal diffusion on mass transfer in a fixed bed while solving the equations. All the solutions and detailed discussions are given by Hines and Maddox[30], Yang[24] and Ruthven[23].

The assumption of isothermal conditions is not valid for a large number of applications if the flow rate is low or the adsorbate has a large heat of adsorption, such as during air dehumidification by desiccants. Local temperature rise due to the heat of adsorption can reduce the adsorption capacity significantly. Derr[60] and Getty and Armstrong[61] reported high temperature rise during gas drying using activated alumina. Hines and Ghosh[11] also observed similar phenomenon while dehumidifying air using silica gel and a molecular sieve. Therefore, to accurately predict the adsorber behavior, the energy and mass transfer equations should be solved simultaneously. The energy balance equation for a packed bed adsorber assuming that local equilibrium is instantaneous between the fluid and particles is given by Cen and Yang[62].

$$\frac{\partial(v\,C\,C_p\,T)}{\partial z} + \varepsilon\,\frac{\partial(C\,C_p\,T)}{\partial t}$$

$$+ \frac{\rho}{V_m}\,\frac{\partial[Q\,(C_{pa}\,T - H)]}{\partial t} + \rho\,C_{ps}\,\frac{\partial T}{\partial t} + \frac{2\,h}{r}\,(T - T_0) = 0 \qquad (10\text{-}27)$$

where

v: superficial flow rate,
C: total concentration of all the contaminants in the gas phase,
C_p: heat capacity of the gas phase,
C_{pa}: heat capacity of the adsorbed phase,
C_{ps}: heat capacity of adsorbent,
T: temperature,
z: length of the column,
V_m: molar volume at the standard temperature and pressure,
Q: adsorbed volume,
H: heat of adsorption,
ρ: bed density,
ε: bed void fraction,
h: overall heat transfer coefficient,
r: radius of the column, and
T_0:   ambient temperature.

The last term in Eqn. (10-26) represents the heat transfer with the surroundings. For a multicomponent system,

$$H = \sum x_i H_i \tag{10-28}$$

$$C_p = \sum y_i C_{pi} \tag{10-29}$$

$$C_{pa} = \sum x_i C_{pi} \tag{10-30}$$

A rigorous analysis for adsorption-desorption involving two adsorbates using the Langmuir isotherm was provided by Glueckauf[63]. In general, calculations for a multicomponent adsorption system are highly complicated, and an analytical solution is not available. For an N-component mixture, N mass balance equations (Eqn. 10-24), one heat transfer equation (Eqn. 10-27), N mass transfer rate equations (Eqn. 10-26), and N equilibrium isotherm equations (Eqn. 10-25) are to be solved simultaneously with proper initial and boundary conditions. Both the extended Langmuir isotherm and the LRC are noniterative and, therefore, are used extensively in the numerical calculations. Recently, the IAS theory based models, although they are iterative, are increasingly used in such calculations because of the better accuracy. The availability of modern high-speed computers reduces the computation time considerably for the iterative process.

Basically three types of numerical methods have been used for solving the above equations. They are the method of characteristics, the orthogonal collocation method, and the finite difference method. The method of characteristics and orthogonal method are recommended for uniform initial conditions, such as in the design of a carbon adsorber. However, for cyclic processes with varying boundary and initial conditions, such as desiccant dehumidifier, it is more convenient to use finite difference method.

10.7 REFERENCES

[1]Amazeen, J., "Desiccant Humidity Control System," AiResearch Mfg., Co., Torrance, CA, NTIS Document No. N74-12499/IGI, 1973.

[2]Lunde, P.J., and F.L. Kester, "Desiccant Humidity Control System," Report No. NASA-CR-115568, 1975.

[3]Relwani, S.M., and D.J. Moschandreas, "Indoor Pollution Control Capabilities of a Desiccant Dehumidifier System," Gas Research Institute, Chicago, IL., Report No. GRI 86/0200, 1986.

[4]Relwani, S.M., D.J. Moschandreas, and I.H. Billick, "Indoor Air Quality Control Capabilities of Desiccant Materials," Proceedings of the 4th International Conference on Indoor Air Quality and Climate, West Berlin, Germany, Vol. 3, pp. 236-240, 1987.

[5]Novosel, D., S.M. Relwani, and D.J. Moschandreas, "Development of a Desiccant Based Environmental Control Unit," Proceedings of IAQ '87: Practical Control of Indoor Air Problems, American Society of Heating, Ventilating and Air-Conditioning Engineers, Inc., Atlanta, GA, pp. 261-270, 1987.

[6]Novosel, D., D.H. McFadden, and S.M. Relwani, "Desiccant Air Conditioning To Control IAQ In Residences," Proceedings of IAQ '88: Engineering Solution to Indoor Air Problems, American Society of Heating, Ventilating and Air-Conditioning Engineers, Inc., Atlanta, GA, pp. 148-165, 1988.

[7]Kuo, S.L., E.O. Pedram, and A.L. Hines, "Analysis on Ammonia Adsorption on Silica Gel Using the Modified Potential Theory," *J. Chem. Eng. Data*, Vol. 30, pp. 330-332, 1985.

[8]Kuo, S.L., and A.L. Hines, Adsorption of Chlorinated Hydrocarbon Contaminants on Silica Gel," *Sep. Sci. Technol.*, Vol. 23(4&5), pp. 273-303, 1988.

[9]Hines, A.L., S.L. Kuo, and N. Dural, "A New Isotherm for Adsorption on Heterogeneous Adsorbents," *Sep. Sci. Technol.*, Vol. 25(7), pp. 869-888, 1990.

[10]Ghosh, T.K., and A.L. Hines, "Adsorption of Acetaldehyde, Propionaldehyde and Butyraldehyde on Silica Gel," *Sep. Sci. Technol.*, Vol. 25(11&12), pp. 1101-1115, 1990.

[11]Hines, A.L., and T.K. Ghosh, "Water Vapor Uptake and Removal of Chemical Contaminants by Solid Adsorbents," Gas Research Institute, Chicago, IL, Report No. GRI- 92/0157.2, 1993.

[12]Okazaki, M., H. Tamon, and R. Toei, "Prediction of Binary Adsorption Equilibria of Solvent and Water Vapor on Activated Carbon," *J. Chem. Eng.* (Japan), Vol. 11(3), pp. 209-215, 1978.

[13]Hassan, N.M., T.K. Ghosh, A.L. Hines, and S.K. Loyalka, "Water Vapor Adsorption on BPL Activated Carbon," *Carbon*, Vol. 29, pp. 681-684, 1991.

[14]Hobbs, F.D., C.S. Parmele, and D.A. Barton, "Survey of Industrial Appli-

cations of Vapor-Phase Activated-Carbon Adsorption For Control of Pollutant Compounds From Manufacture of Organic Compounds," Report No. EPA-600/Z-83-035, 1983.

[15]Hines, A.L., T.K. Ghosh, S.K. Loyalka, and R.C. Warder, "Investigation of Co-Sorption of Gases And Vapors as a Means to Enhance Indoor Air Quality," Gas Research Institute, Chicago, GRI-90/01934, NTIS Document No. PB91-178806/GAR, 1990.

[16]Nelson, G.O., and A.M. Correia, "Respirator Cartridge Efficiency Studies, VII-Summary and Conclusions," *Am. Ind. Hyg. Assoc. J.*, Vol. 37, pp. 514-525, 1976.

[17]Reucroft, P.J., H.K. Patel, W.C. Russell, and W.M. Kim, "Modeling of Equilibrium Gas Adsorption for Multicomponent Vapor Mixtures, Part II," NTIS Document No. AD-A174058, 1986.

[18]Hines, A.L., T.K. Ghosh, S.K. Loyalka, and R.C. Warder, *Indoor Air : Quality and Control*, Prentice Hall, Englewood Cliffs, NJ, 1993.

[19]Brunauer, S., *The Adsorption Of Gases And Vapors*, Princeton University Press, Princeton, NJ, 1945.

[20]Brunauer, S., P.H., Emmett, and E. Teller, "Adsorption of Gases in Multimolecular Layers," *J. Am. Chem. Soc.*, Vol. 60, pp. 309-319, 1938.

[21]Young, D.M., and Crowell, A.D., *Physical Adsorption Of Gases*, Butterworth, London, 1962.

[22]Ross, S., and J.P. Olivier, *On Physical Adsorption*, Wiley-Interscience, New York, NY, 1964.

[23]Ruthven, D.M., *Principles Of Adsorption And Adsorption Processes*, John Wiley & Sons, Inc., New York, 1984.

[24]Yang, R.T., *Gas Separation By Adsorption Processes*, Butterworth Publishers, Boston, MA, 1987.

[25]Langmuir, I, "The Adsorption of Gases on Plane Surfaces of Glass, Mica, and Platinum," *J. Am. Chem. Soc.*, Vol. 40, pp. 1361-1403, 1918.

[26]Zel'dovich, Y., "Theory of the Freundlich Adsorption Isotherm," *Acta Physicochim*, U.R.S.S., Vol. 1, pp. 961-974, 1934.

[27]Jovanovic, D.S., "Physical Adsorption of Gases. I. Isotherms for Monolayer and Multilayer Adsorption," *Kolloid-Z.Z. Polym.*, Vol. 235, pp. 1203-1213, 1969.

[28]Sircar, S., "New Adsorption Isotherm for Energetically Heterogeneous Adsorbents," *J. Colloid Interface Sci.*, Vol. 98, pp. 306-318, 1984a.

[29]Sircar, S., "Effect of Local Isotherm on Adsorbent Heterogeneity," *J. Colloid Interface Sci.*, Vol. 101, pp. 452-461, 1984b.

[30]Hines, A.L., and R.N. Maddox, *Mass Transfer: Fundamentals And Applica-*

tions, Prentice Hall, NJ, 1985.

[31]Ghosh, T.K., "Adsorption of Acetaldehyde, Propionaldehyde, and Butyraldehyde on Silica Gel and Molecular Sieve-13X", Doctorate Dissertation, Oklahoma State University, 1989.

[32]Suwanayuen, S., and R.P. Danner, "A Gas Adsorption Isotherm Equation Based on Vacancy Solution Theory," *AIChE J.*, Vol. 26, pp. 68-76, 1980a.

[33]Suwanayuen, S., and R.P. Danner, "Vacancy Solution Theory of Adsorption from Gas Mixtures," *AIChE J.*, Vol. 26, pp. 76-83 1980b.

[34]Cochran, T.W., R.L. Kabel, and R.P. Danner, "Vacancy Solution Theory of Adsorption Using Flory-Huggins Activity Coefficient Equations," *AIChE J.*, Vol. 31, PP. 268-277, 1985.

[35]Markham, E.C., and E.F. Benton, "The Adsorption of Gas Mixture by Silica," *J. Am. Chem. Soc.*, Vol. 53, pp. 497-507, 1931.

[36]Broughton, D.B., "Adsorption Isotherms for Binary Mixtures," *Ind. Eng. Chem.*, Vol. 40, pp. 1506-1508, 1948.

[37]Kemball, C., E.K. Rideal, and E.A. Guggenheim, "Thermodynamics of Monolayer," *Trans. Faraday Soc.*, Vol. 44, pp. 948-954, 1948.

[38]Innes, W.B., and H.H. Rowley, "Adsorption Isotherms of Mixed Vapors of Carbon Tetrachloride and Methanol on Activated Charcoal at 25°C," *J. Phys. & Colloid Chem.*, Vol. 51, pp. 1154-1171, 1947.

[39]Sips, R., "Structure of Catalyst Surface," *J. Chem. Phys.*, Vol. 16, pp. 490-495, 1948.

[40]Koble, R.A., and T.E. Corrigan, "Adsorption Isotherms for Pure Hydrocarbons," *Ind. Eng. Chem.*, Vol. 44, pp. 383-387, 1952.

[41]Yon, C.M., and P.H. Turnock, "Multicomponent Adsorption Equilibria on Molecular Sieves," *AIChE Symposium Series*, Vol. 67, No. 117, pp. 73-83, 1971.

[42]Arnold, J.R., "Adsorption of Gas Mixtures, Nitrogen-Oxygen on Anatase," *J. Am. Chem. Soc.*, Vol. 71, pp. 104-110, 1949.

[43]Lewis, W.K., E.R. Gilliland, B. Chertow, and W.P. Cadogan, "Adsorption Equilibria: Hydrocarbon Gas Mixtures," *Ind. Eng. Chem.*, Vol. 42, pp. 1319-1326, 1950.

[44]Myers, A.L., and J.M. Prausnitz, "Thermodynamics of Mixed Gas Adsorption," *AIChE J.*, Vol. 11, pp. 121-127, 1965.

[45]Cook, W.H., and D. Basmadjian, "The Prediction of Binary Adsorption Equilibria from Pure Component Isotherms," *Can. J. Chem. Eng.*, Vol. 43, pp. 78-83, 1965.

[46]Sircar, S., and A.L. Myers, "Surface Potential Theory of Multilayer Ad-

sorption from Gas Mixtures," *Chem. Eng. Sci.*, Vol. 28, pp. 489-499, 1973.

[47]O'Brien, J.A., and A.L. Myers, "Rapid Calculation of Multicomponent Adsorption Equilibria from Pure Isotherm Data," *Ind. Eng. Chem. Process Des. Dev.*, Vol. 24, pp. 1188-1191, 1985.

[48]Moon, H., and C. Tien, "Further Work on Multicomponent Adsorption Equilibria Calculations Based on the Ideal Adsorbed Solution Theory," *Ind. Eng. Chem. Res.*, Vol. 26, pp. 2042-2047, 1987.

[49]Costa, E., J.L. Sotelo, G. Calleja, and C. Marron, "Adsorption of Binary and Ternary Hydrocarbon Gas Mixtures on Activated Carbon: Experimental Determination and Theoretical Prediction of the Ternary Equilibrium Data," *AIChE J.*, Vol. 27, pp. 5-12, 1981.

[50]Hyun, S.H., and R.P. Danner, "Equilibrium Adsorption of Ethane, Ethylene, Isobutane, Carbon Dioxide and Their Binary Mixtures on 13X Molecular Sieve," *J. Chem. Eng. Data*, Vol. 27, pp. 196-200, 1982.

[51]Talu, O., and I. Zwiebel, "Multicomponent Adsorption Equilibria of Nonideal Mixtures," *AIChE J.*, Vol. 32(8), pp. 1263-1276, 1986.

[52]Ghosh, T.K., Hon-Da Lin, and A.L. Hines, "Hybrid Adsorption-Distillation Process for Separating Propane and Propylene," *Ind. Eng. Chem. Res.*, Vol. 32(10), pp. 2390-2399, 1993.

[53]Jaroniec, M., "Description of Kinetics and Equilibrium State of Adsorption from Multicomponent Gas Mixtures on Solid Surfaces," *Thin Solid Films*, Vol. 71, pp. 273-304, 1980.

[54]Hougen, O.A., and W.R. Marshall, "Adsorption from a Fluid Stream Flowing Through a Stationary Granular Bed," *Chem. Eng. Prog.*, Vol. 43, pp. 197-208, 1947.

[55]Rosen, J.B., "Kinetics of a Fixed Bed System for Solid Diffusion into Spherical Particles," *J. Chem. Phys.*, Vol. 20, pp. 387-394, 1952.

[56]Rosen, J.B., "General Numerical Solution for Solid Diffusion in Fixed Beds," *Ind. Eng. Chem.*, Vol. 46, pp. 1590-1594, 1954.

[57]Hiester, N.K., and T. Vermeulen, "Saturation Performance of Ion Exchange and Adsorption Columns," *Chem. Eng. Prog.*, Vol. 48, pp. 505-516, 1952.

[58]Van Deemter, J.J., F.J. Zuiderweg, and A. Klinkenberg, "Longitudinal Diffusion and Resistance to Mass Transfer as Causes of Nonideality in Chromatography," *Chem. Eng. Sci.*, Vol. 5, pp. 271-287, 1956.

[59]Acrivos, A., "Combined Effect of Longitudinal Diffusion and External Mass Transfer Resistance in Fixed Bed Operation," *Chem. Eng. Sci.*, Vol. 13, pp. 1-6, 1960.

[60]Derr, R.B., "Drying Air and Commercial Gases with Activated Alumina," *Ind. Eng. Chem.*, Vol. 30, pp. 384-388, 1938.

[61]Getty, R.J., and W.P. Armstrong, "Drying Air with Activated Alumina Under Adiabatic Conditions," *Ind. Eng. Chem. Proc. Des. Dev.*, Vol. 3, pp. 60-66, 1964.

[62]Cen, P., and R.T. Yang, "Separation of Five-Component Gas Mixture by Pressure Swing Adsorption," *Sep. Sci., Technol.*, Vol. 20(9&10), pp. 725-747, 1985.

[63]Glueckauf, E., "Theory of Chromatography-VII, The General Theory of Two Solutes Following Nonlinear Isotherms," *Discussions Faraday Soc.*, No. 7, pp. 12-25, 1949.

[64]Yeh, R-L., "Adsorption of Water Vapor, Toluene, 1,1,1-Trichloroethane, and Carbon Dioxide on Silica gel, Molecular Sieve 13X, and Activated Carbon," Master of Science Thesis, University of Missouri-Columbia, 1991.

[65]Hassan, N.M., "Co-Adsorption of Radon and Water-Vapor Mixtures on BPL Activated Carbon and Solid Desiccants," Doctorate Dissertation, University of Missouri-Columbia, 1990.

11

Desiccant-Assisted Package and Outside Air Preconditioning Modules

Milton Meckler, P.E.
President, The Meckler Group
Encino, California

11.1 INTRODUCTION

Airborne microorganisms, which are responsible for many acute diseases, infections and allergies, are well protected indoors by the moisture surrounding them. While the human body is generally the host for various bacteria and viruses, fungi can grow in moist places. It has been concluded that an optimum relative humidity (RH) range of 40% to 60% is necessary to minimize or eliminate the bacterial, viral and fungal growth. In addition, humidity also has an effect on air cleanliness. It reduces the presence of dust particles and prevents the deterioration of the building structure and its contents. Therefore, controlling humidity is a very important factor to human comfort in minimizing adverse health effects and maximizing the structural longevity of the building.

In recent years, liquid and solid desiccant applications have been explored with respect to independent humidity control[1,2]. A great deal of research[3-6] has been done, and is continuing, in the use of solid desiccant systems including innovative regeneration methods. Hybrid systems incorporating both desiccant and conventional refrigeration cycles integrated within a single enclosure[7] are now being actively studied. These systems offer the advantage of substantially enhancing indoor air quality (IAQ) by simultaneously removing moisture and contaminants from indoor air by co-sorption.

In this chapter, we will introduce two proprietary desiccant-assisted air conditioning systems. They both combine hybrid dehumidification and mechanical refrigeration through a desiccant-preconditioning module to precondition the minimum outside air. Both systems also employ indirect evaporative cooling within the desiccant-preconditioning module. One of these systems is the Desiccant-Assisted Rooftop Package Air-Conditioning (DRPA/C) system[7] that employs a solid desiccant wheel/heat-pipe configuration within the desiccant-preconditioning module. In this system, dehumidification and mechanical refrigeration are combined in a single package unit. The second system is the Desiccant-Assisted Distributed Rooftop Air-Conditioning (DDRA/C) system[8] which employs a desiccant wheel/compact indirect heat-exchanger (e.g., sensible heatwheel). This system has a single desiccant-preconditioning module that serves two or more conventional rooftop air-conditioning (A/C) units.

Use of the proposed desiccant-preconditioning module can improve the humidity control capabilities of overall refrigeration systems due to inherent evaporator coil limitations particularly for southeastern U.S. climates. The desiccant-preconditioning module can more efficiently remove the sensible and latent cooling loads than a conventional A/C unit. Furthermore, incorporation of desiccant-preconditioning into such systems can accommodate the higher minimum outside air ventilation rates (i.e., 20 cubic feet per minute [cfm] for office occupancies) now required by ASHRAE Standard 62-1989. This is a major problem associated with the conventional Rooftop Package Air-Conditioning (RPA/C) systems that need to satisfy the energy guidelines set by the American Society of Heating, Refrigerating and Air-Conditioning Engineers, Inc. (ASHRAE) and Air-Conditioning and Refrigeration Institute (ARI).

In addition to introducing the two systems mentioned above, the results of a comparative study between the DDRA/C system and a conventional RPA/C system for an 11,000-ft² nominal, one-story, prototype office building sited in five U.S. cities with varying microclimates, will be presented. First, we will explore the importance of outside air humidity control and the role of indirect evaporative cooling.

11.2 HUMIDITY CONTROL/INDIRECT EVAPORATIVE COOLING

Figure 11-1 shows the total load required to cool 1000 cfm of outside air from various wet-bulb temperatures to a room condition of 75°F dry-bulb temperature at various RH. Referring to Figure 11-1, notice that if an indoor condition (75°F dry-bulb temperature and 50% RH) is to be maintained, approximately five tons of mechanical refrigeration cooling is required for the total load removal associated with each 1000 cfm of outside air treated at an approximately 78°F wet-bulb temperature[9]. Clearly, this can be substantially reduced by means of desiccant pretreatment. Integrating desiccant cooling capability within the preconditioning module produces air delivery to the downstream indoor fan coil unit that can result in substantially lower initial cost and annual energy savings. This results from convenient adjacencies with other airstreams (i.e., return air) capable of extracting further cooling potential. The ventilation refrigeration load as represented in Figure 11-1 can be substantially reduced by harnessing the (already dehumidified) building exhaust air's remaining cooling potential prior to its being expelled outdoors.

Indirect evaporative cooling[10] represents a significant advance in the use of heat-pipes. A heat-pipe is basically a tube that is fabricated with a capillary wick structure. It is evacuated, filled with refrigerant and permanently sealed (refer to Figure 11-2). Thermal energy applied to either end of the heat-pipe causes the refrigerant at that end to vaporize. The vapor then travels to the other end of the heat-pipe where the thermal energy is removed. As that happens, the vapor condenses back to liquid, giving up the latent heat of condensation. The liquid flows back to the other end of the heat-pipe to be reused, thus completing the cycle.

Figure 11-3 shows a heat-pipe module with evaporative cooling. In Figure 11-3, notice that air passing through the exhaust-side of the heat-pipe is sprayed and washed with water. Evaporation occurs on the fins of the heat-pipe, raising the moisture content of only the exhaust air. The

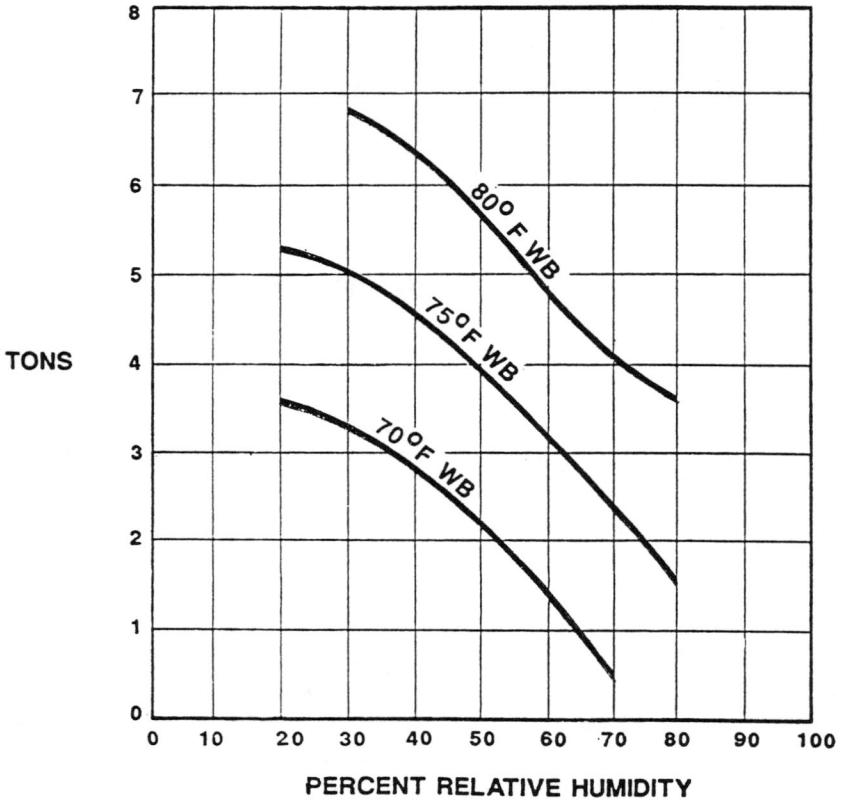

PERCENT RELATIVE HUMIDITY

Reference : Meckler, M. "Integrated Desiccant Cold Air Distribution,"
Heating/Piping/Air Conditioning, May 1989.

Figure 11-1. Total Load Required to Cool 1000 cfm of Outside Air from Various Wet-Bulb Temperatures to a Room Condition of 75°F Dry-Bulb at Various Relative Humidities.

heat of vaporization needed to evaporate the water on the exhaust-side comes from the supply-side fin surfaces, cooling the incoming air. The result is a cycle in which supply air is constantly dry-cooled, thereby reducing the size of the downstream mechanical air-conditioning equipment and allowing significant reductions in electrical power consumption.

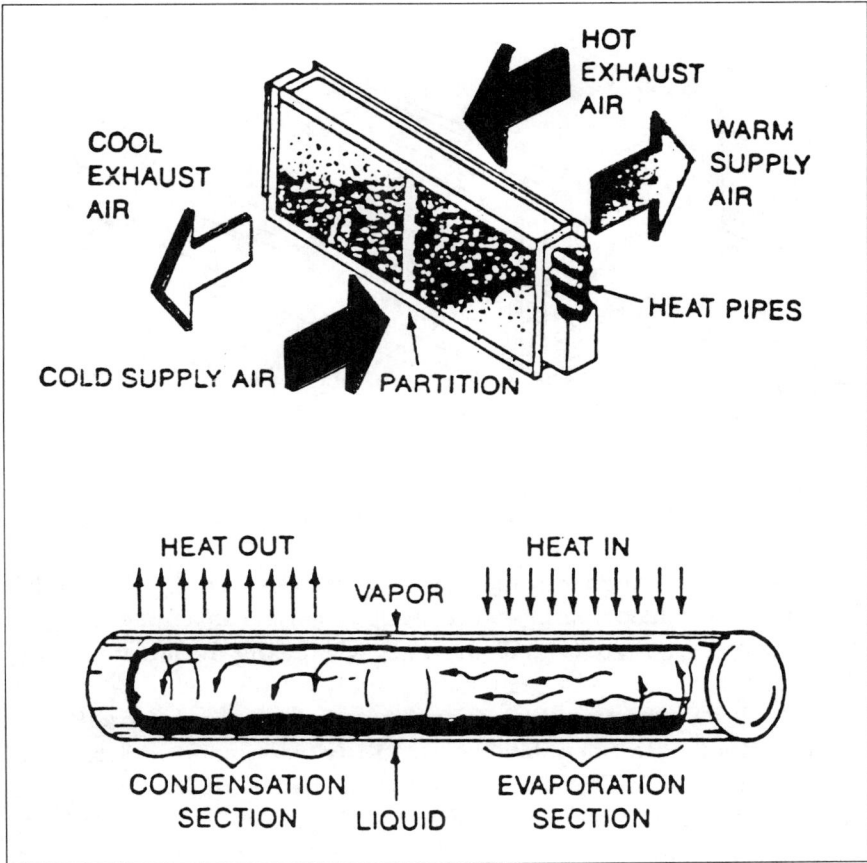

Figure 11-2. A Basic Heat-Pipe.

11.3 DESICCANT-ASSISTED ROOFTOP PACKAGE AIR-CONDITIONING SYSTEM

The DRPA/C system shown in Figure 11-4 employs electrically powered mechanical refrigeration sensible cooling combined with desiccant dehumidification, and efficiently removes the sensible and latent cooling loads. As shown in Figure 11-4, the desiccant-preconditioning module in this hybrid system is integrated into a conventional rooftop A/C unit with an outdoor air/return air mixing plenum allowing for incorporation of an economizer cycle. The downstream vapor-compression

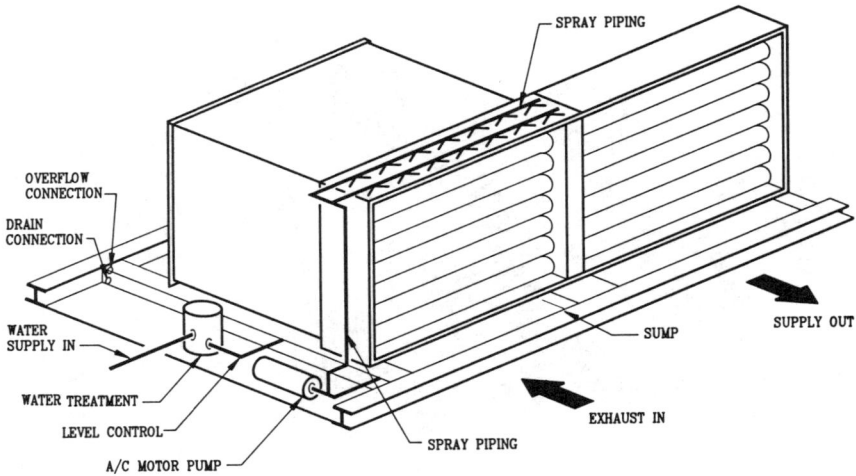

Figure 11-3. A Heat-Pipe Module with Evaporative Cooling.

system operates at higher evaporator temperatures resulting in a higher thermal coefficient of performance (COP) than a conventional vapor-compression system (refer to Figure 11-5) where both the latent and sensible loads are removed by a common evaporator coil.

The desiccant-preconditioning module operates to remove sufficient moisture from outside ventilation air to satisfy the combined internal and external latent loads. The remaining sensible load is satisfied by smaller vapor-compression system components. Consequently, this hybrid system requires much less energy because of an increased overall system COP. Heat normally rejected to ambient by the refrigeration condenser air coil is utilized to regenerate the desiccant, eliminating the need for external regenerative heat at or below 135°F with reciprocating compressors or at a substantially higher temperature with rotary or scroll displacement compressors.

11.3.1   Desiccant Dehumidification

Supply air to a zone consists of a mixture of outside air and return air that has been conditioned. The minimum outside air is dehumidified and mixed with return air and controlled by a damper prior to cooling by means of direct expansion (DX) in evaporator coils of the refrigeration system and discharged into the zone. Approximately two-thirds of the free area of the desiccant rotary bed is exposed to the outside air duct,

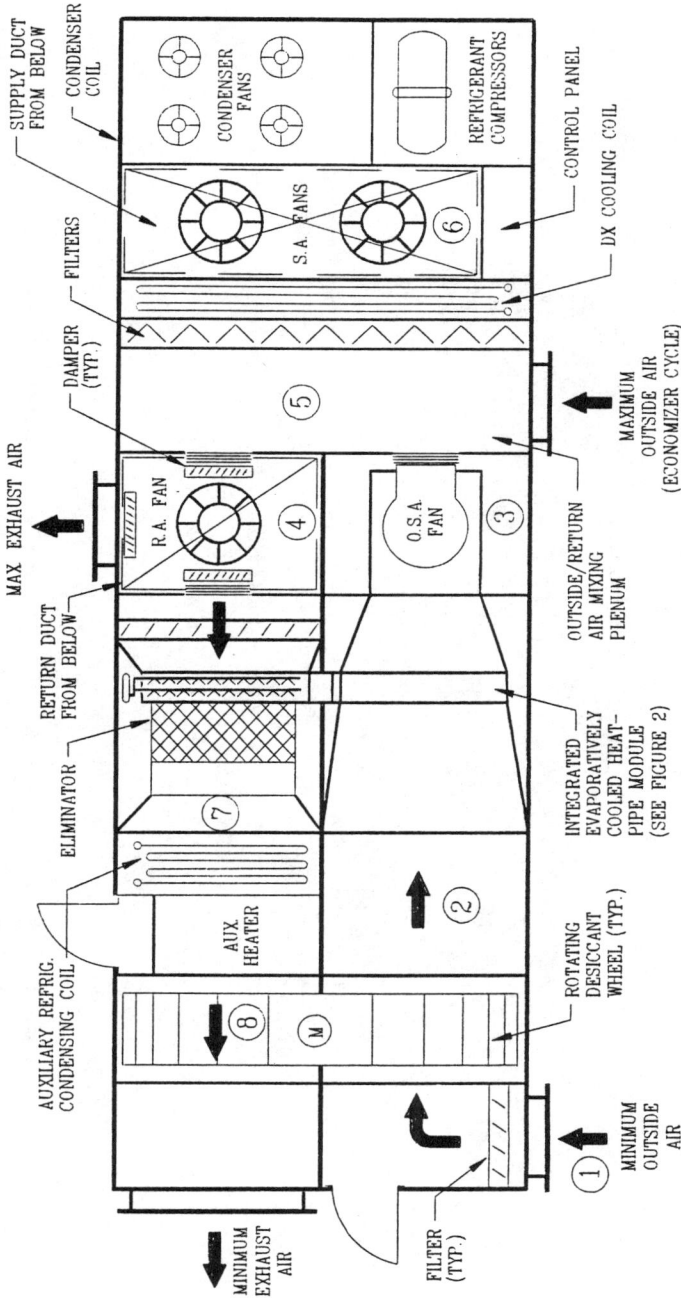

Figure 11-4. Desiccant-Assisted Rooftop Package Air-Conditioning System.

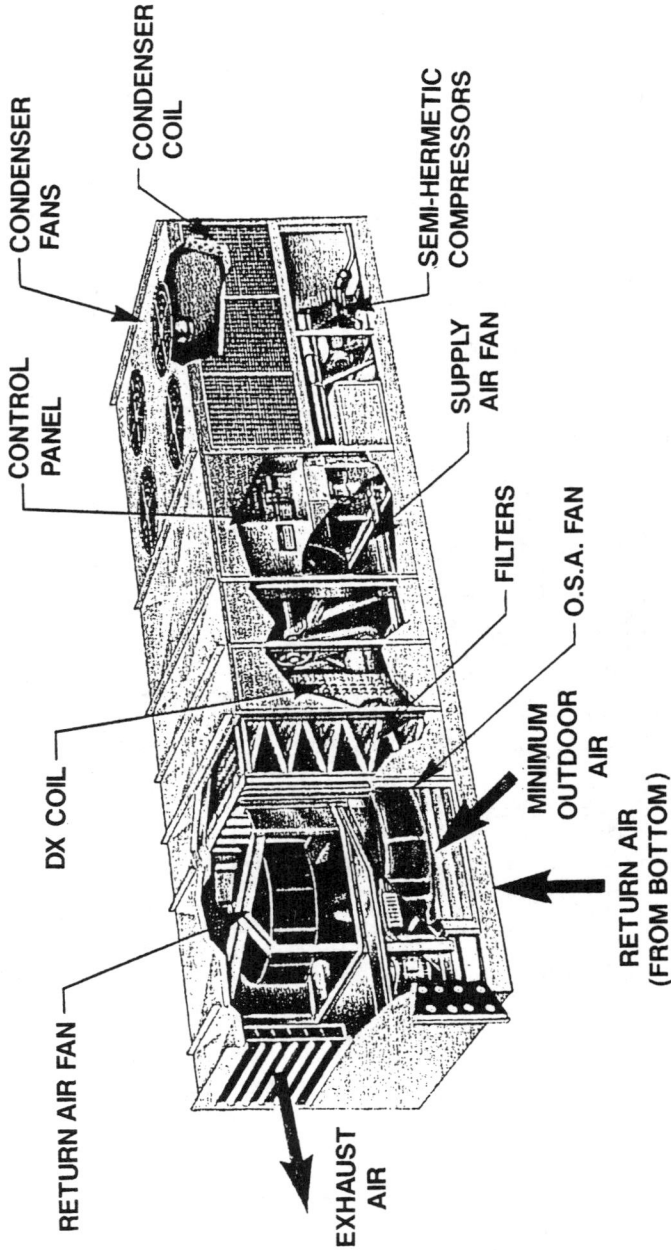

Figure 11-5. A Conventional Rooftop Package Air-Conditioning System.

while approximately one-third is exposed to the regenerating air duct. Regeneration air heated by an auxiliary condenser coil as shown in Figure 11-4 is then discharged to the atmosphere following regeneration of the desiccant in the manner shown. The air ducts are sealed with the rotary bed of the desiccant rotated slowly by means of a variable-speed motor.

The solid-desiccant system shown in Figure 11-4 employs a rotary wheel which is comprised of a series of tubes or plates of a solid desiccant to which air is continuously exposed. When air is dried with a desiccant, its temperature rises because the latent heat and heat from the regenerated desiccant are transferred to the dried air. This temperature-rise is compensated by providing after cooling associated with the exhaust of excess return air to atmosphere (utilizing an evaporative cooling effect to remove heat leaving the dehumidifying desiccant by heat-pipe means as subsequently described) and reduces the load otherwise imposed on the downstream refrigerant evaporator coil.

The performance of a desiccant-preconditioning module having a solid desiccant wheel is a function of the desiccant material, internal geometry of the module, and the equipment operating parameters. The material type affects the size, range of operation (temperature and humidity), efficiency, cost and the service life. The choice of desiccant also affects the thermal COP and cooling capacity of the system. The geometry of a module affects its pressure-drop, size, and the cost, thus the thermal and electrical COPs and cost of the cooling system.

Control strategies also have an affect on the overall system performance. Silica gel has a high moisture recycling capacity, whereas lithium chloride (LiCl), a hygroscopic salt used in some available commercial dehumidifiers, is employed in selected applications where saturated air conditions are not anticipated. Other materials include hygroscopic polymers, zeolites and alumina. Parallel-passage geometries have high heat- and mass-transfer rates and a low pressure-drop. Savings in refrigeration capacity can be achieved by using a solid desiccant wheel instead of a conventional cooling coil to remove the latent load associated with minimum outside ventilation air flow. Accordingly, minimum outside air requirements necessary for maintaining acceptable IAQ increase significantly because of tighter building envelopes, and greater IAQ concerns.

11.3.2   Effect of Liquid Cylinder Cooling

It is possible to compress an ideal gas isothermally with less power than isentropically. A polytropic compression falls somewhere between these two extremes approaching that of isentropic compression as "n" approaches "k" and approaching isothermal compression as "n" approaches 1. Figure 11-6 shows these compression processes on a P-V diagram.

The isentropic compression work is determined by the following expression:

$$W = [(kp_1V_1)/(1-k)][(P_2/P_1)^{(k-1)/k} - 1] \qquad (11-1)$$

If "n" is substituted for "k," Eqn. (11-1) becomes the expression for the polytropic compression work. As "n" decreases, the calculated work becomes less. The "n" actually decreases as the cylinder is cooled. As "n"

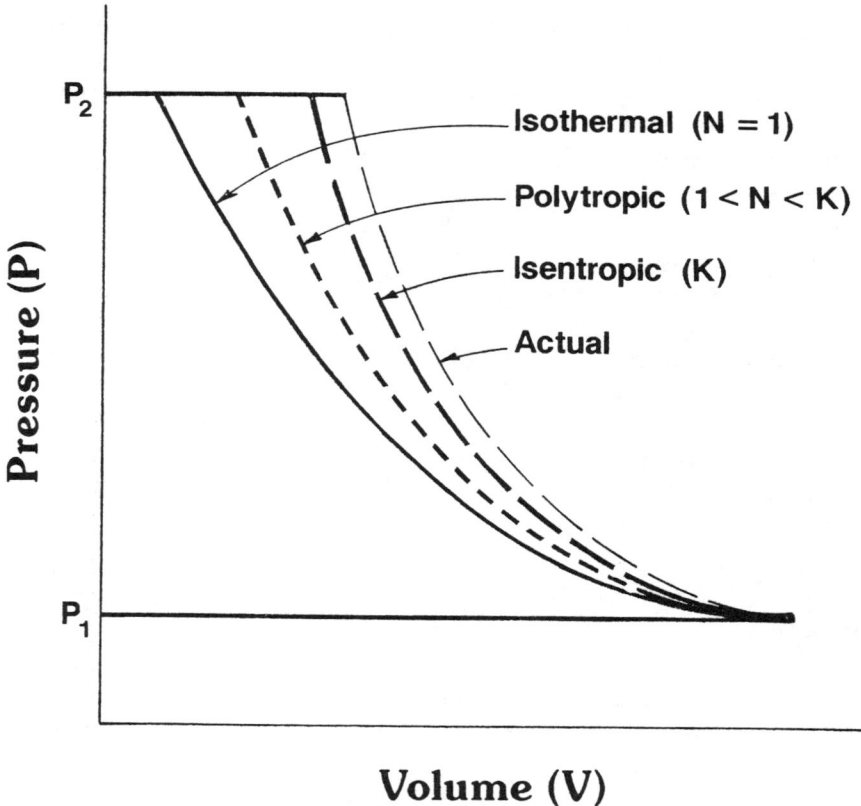

Figure 11-6. Compression Processes.

approaches 1, Eqn. (11-1) becomes indeterminate and is replaced by Eqn. (11-2) below:

$$W = p_1V_1 \ln(V_2/V_1) \tag{11-2}$$

Eqn. (11-2) represents the isothermal compression work. It is theoretically possible for polytropic compression to exist if the gas rejects some heat to the surroundings during the compression process, and the auxiliary refrigerant cooling coil (shown in Figure 11-4) serves this purpose well. When some of the heat is not rejected by the refrigerant gas but instead is added to the gas during the compression process, it results in a compression process with an "n" greater than "k."

The actual compression process is labeled as "Actual" in Figure 11-6 and shows the effect of heat supplied by the motor waste heat and the heat generated by friction. Notice that during a conventional vapor-compression cycle, the shaft-work is normally much greater than the theoretical isentropic compression work. The ratio of this theoretical isentropic work to the actual shaft-work is the isentropic (or reversible-adiabatic) compression efficiency, and it is normally around 65% for most low-temperature applications.

The two most commonly employed choices for auxiliary cooling are desuperheating and liquid injection. Both are equally capable of reducing the discharge-gas temperature to acceptable levels for high compression ratios. The selection of one or the other method or both depends primarily on the application. Liquid injection is simply a means of metering a small amount of liquid from the condenser to the compressor cylinder or compression chamber. Since this injection occurs essentially after the chamber is closed, the capacity is about the same, but the discharge-gas temperature is slightly lower (refer to Table 11-1).

Liquid injection can serve as an effective means to increase the pressure ratio in one stage and to reduce discharge-gas temperature, thereby decreasing motor temperature since compressor motors are cooled by the discharge gas. It is comparable to single-stage cooling used in some screw compressors and can be used in most positive-displacement compressors, including the scroll. For most current air-conditioning applications, the pressure ratio is not sufficient to require any special cooling if the motor is cooled by the suction gas. However, it can be applied to scroll compressors where the pressure ratio is excessively higher.

Table 11-1. Effect of Liquid Injection on Discharge-Gas Temperature.

PARAMETER	INJECTION (% OF CONDENSER FLOW RATE)			
	0%	3.5%	7%	10%
CAPACITY (Ton)	4.6	4.6	4.5	4.5
ENERGY EFFICIENCY RATIO, EER (Btu/W−h)	9.2	9.2	9.0	8.9
DISCHARGE−GAS TEMPERATURE (˚F)	220	204	191	180

Figure 11-7 shows a generalized curve of discharge-gas temperature vs. compression ratio for Refrigerant-22 (R-22). This curve is constructed by using data of many compressors run at a minimum voltage. The compression ratio can be used as an indicator for the need of auxiliary cooling. As the compression ratio increases, the discharge-gas temperature also increases as predicted by the adiabatic constant for R-22. At a maximum system load (e.g., employing rotary compressors), the discharge-gas temperature limit is approximately 240°F. From Figure 11-7, one can see that an R-22 system with compression ratios of 3.5 or greater, is a likely candidate for auxiliary cooling.

Liquid injection has an advantage in that as compression ratio is increased (tending to produce higher discharge-gas temperatures), more liquid will be forced through the capillary, maintaining satisfactory temperature levels. A liquid-injection-cooled compressor utilizes a capillary tube for metering. The capillary should be sized so that the final discharge-gas temperature is in the range of 180°F to 205°F when the system operates at its rated point and below its maximum load condition.

Overinjection of liquid can result in a higher cylinder pressure and, therefore higher power input and reduced capacity; this must be avoided. The injected liquid, exposed to lower pressure, flashes into vapor, cooling the cylinder and discharge gas. This dense, cooler gas passes into the compressor shell, and is brought in contact with the motor to carry off its heat. Correct location of the liquid injection tap on a reverse-cycle system is important to ensure the necessary motor cooling in both the heating and air-conditioning modes (refer to Figure 11-8).

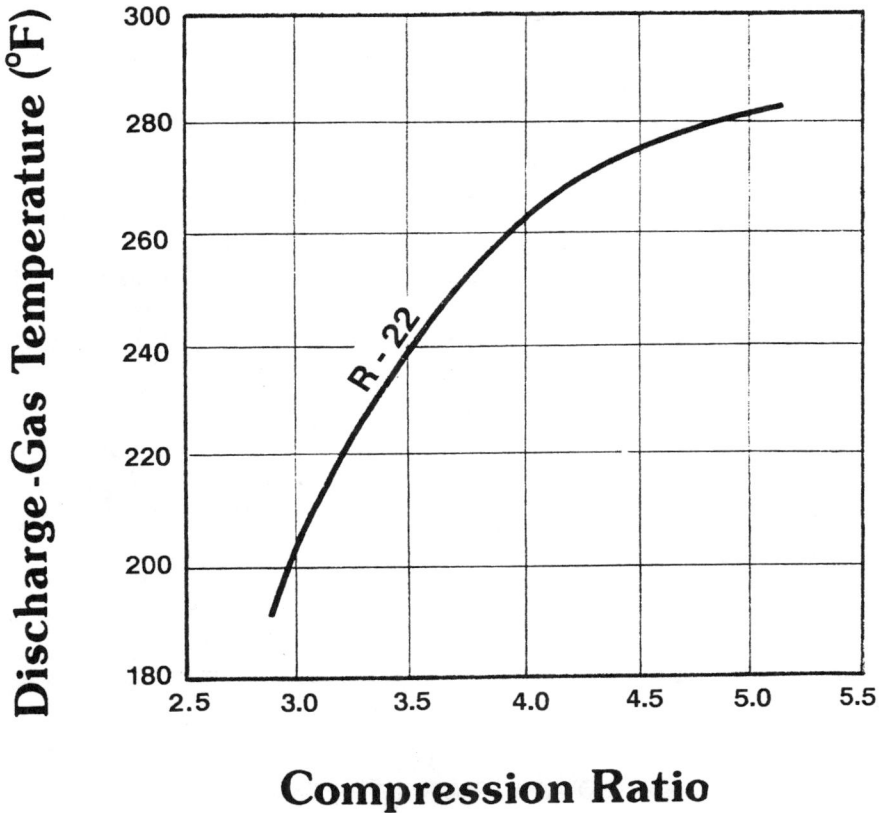

Figure 11-7. Effect of Compression Ratio on Discharge-Gas Temperature for R-22.

Liquid injection will not decrease capacity as a full suction volume will have been taken into the cylinder before the liquid flashes. As one can see from Table 11-1, there is relatively little change (slightly decreased energy efficiency ratio [EER]), except for the gas-discharge temperature. The injected liquid increases the amount of liquid in the system and introduces some system complexity, but it is a fairly simple and effective means to reduce discharge-gas temperature. The advantage of employing liquid injection in modulating regeneration temperatures in response to humidity variations makes up for this minor decrease.

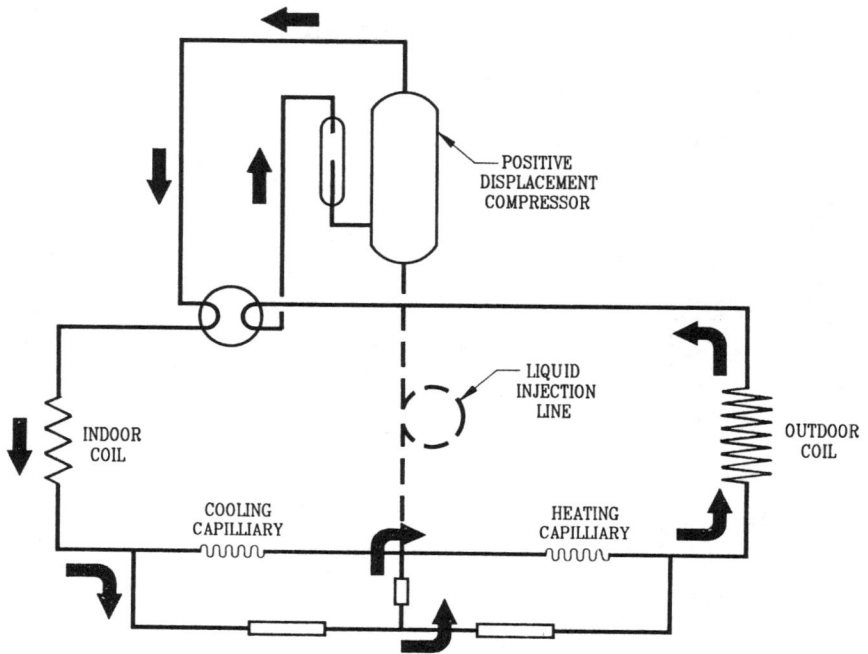

Figure 11-8. Liquid-Line Compressor Tap.

11.4 DESICCANT-ASSISTED DISTRIBUTED ROOFTOP AIR-CONDITIONING SYSTEM

The DDRA/C system also has a desiccant-preconditioning module. As opposed to the module in the DRPA/C system introduced earlier, this module (shown in Figure 11-9) serves two or more conventional A/C units, each with an outdoor air/return air mixing plenum allowing for incorporation of an economizer cycle. In this way, dehumidification and mechanical refrigeration are combined in a split system through a common desiccant-preconditioning module to precondition the outside minimum air (refer to Figure 11-10 showing the airflow diagram for the DDRA/C system). For comparison purposes, an airflow diagram for the conventional RPA/C system is presented in Figure 11-11. The DDRA/C system also employs indirect evaporative cooling within the desiccant-preconditioning module as shown in Figure 11-9.

Figure 11-12 shows the psychrometric chart for the DDRA/C system. Note that all designated points on the psychrometric chart for the

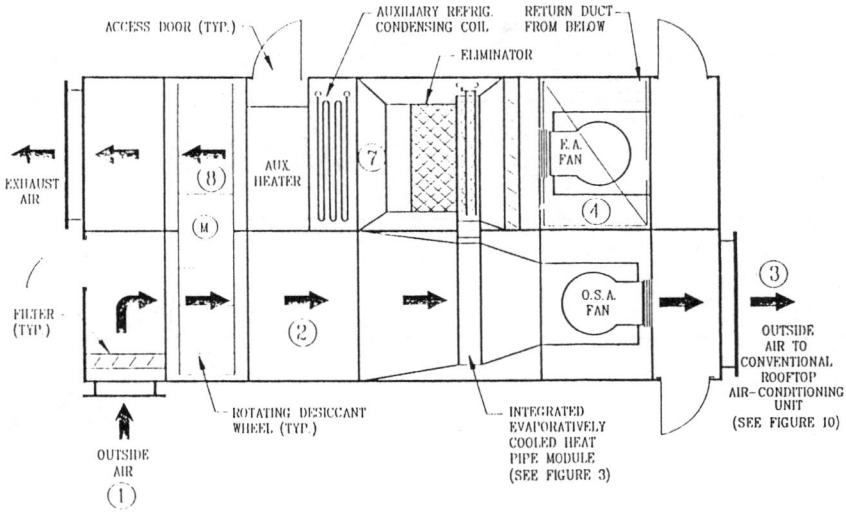

Figure 11-9. Desiccant-Preconditioning Module for DDRA/C System.

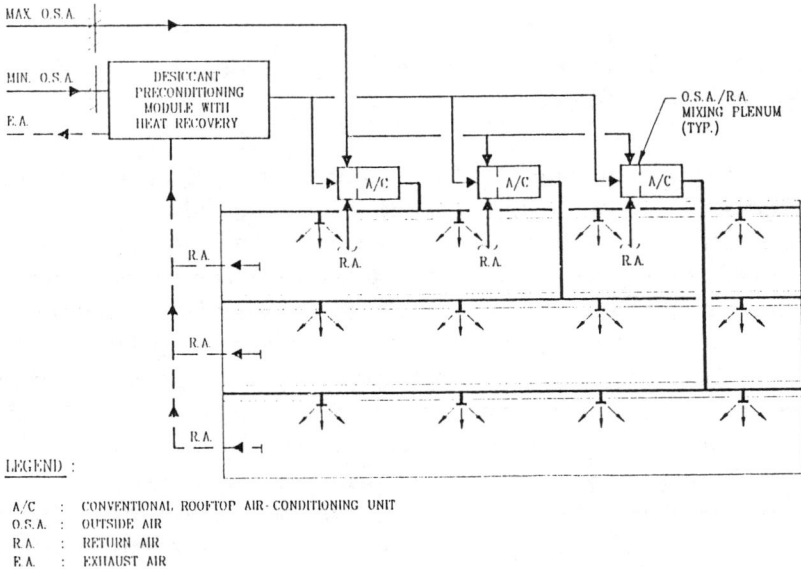

LEGEND :

A/C   :   CONVENTIONAL ROOFTOP AIR-CONDITIONING UNIT
O.S.A.   :   OUTSIDE AIR
R.A.   :   RETURN AIR
E.A.   :   EXHAUST AIR

Figure 11-10. Air Flow Diagram for Desiccant-Assisted Distributed Rooftop Air-Conditioning System.

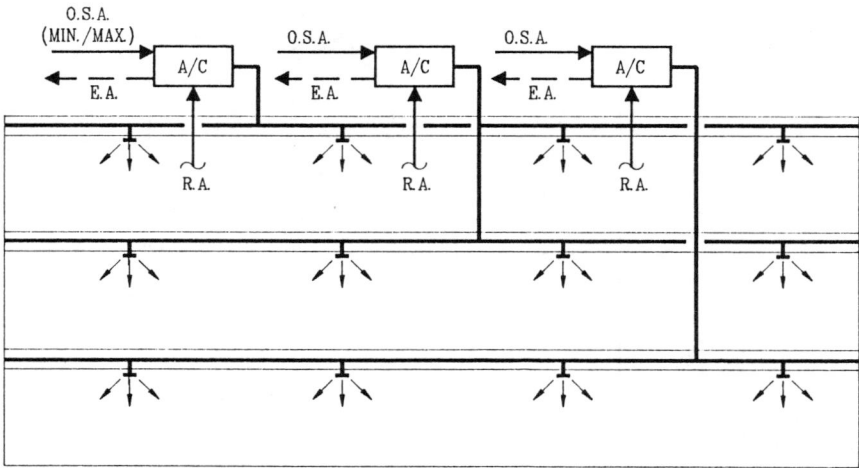

LEGEND :

A/C   :   CONVENTIONAL ROOFTOP AIR-CONDITIONING UNIT
O.S.A.   :   OUTSIDE AIR
R.A.   :   RETURN AIR
E.A.   :   EXHAUST AIR

Figure 11-11. Air Flow Diagram for a Conventional Rooftop Package Air-Conditioning System.

DDRA/C system in Figure 11-12 correspond to the points in Figure 11-9. In Figure 11-12, notice that by incorporating evaporative cooling sprays for exhaust air out of the building, heat is removed from adjacent incoming outside air (into the building as ventilation air). This provides substantial cooling effect (on summer cooling) by rejecting heat removed from outside air to the heat sink created by counterflowing evaporatively cooled exhaust air. A similar benefit also takes place on winter heating when exhaust air sprays are inoperative and warm exhaust air is used as the heat source for preheating incoming cold outside ventilation air.

Referring to Figure 11-9, notice that after the exhaust air removes sensible and latent heat transferred from outside air in summer cooling (refer to Figure 11-12), it passes through the auxiliary refrigerant condenser coil which further preheats exhaust air. This refrigerant condenser coil simultaneously serves as a refrigerant desuperheater (resulting in an improved refrigeration-cycle efficiency) while providing supplementary

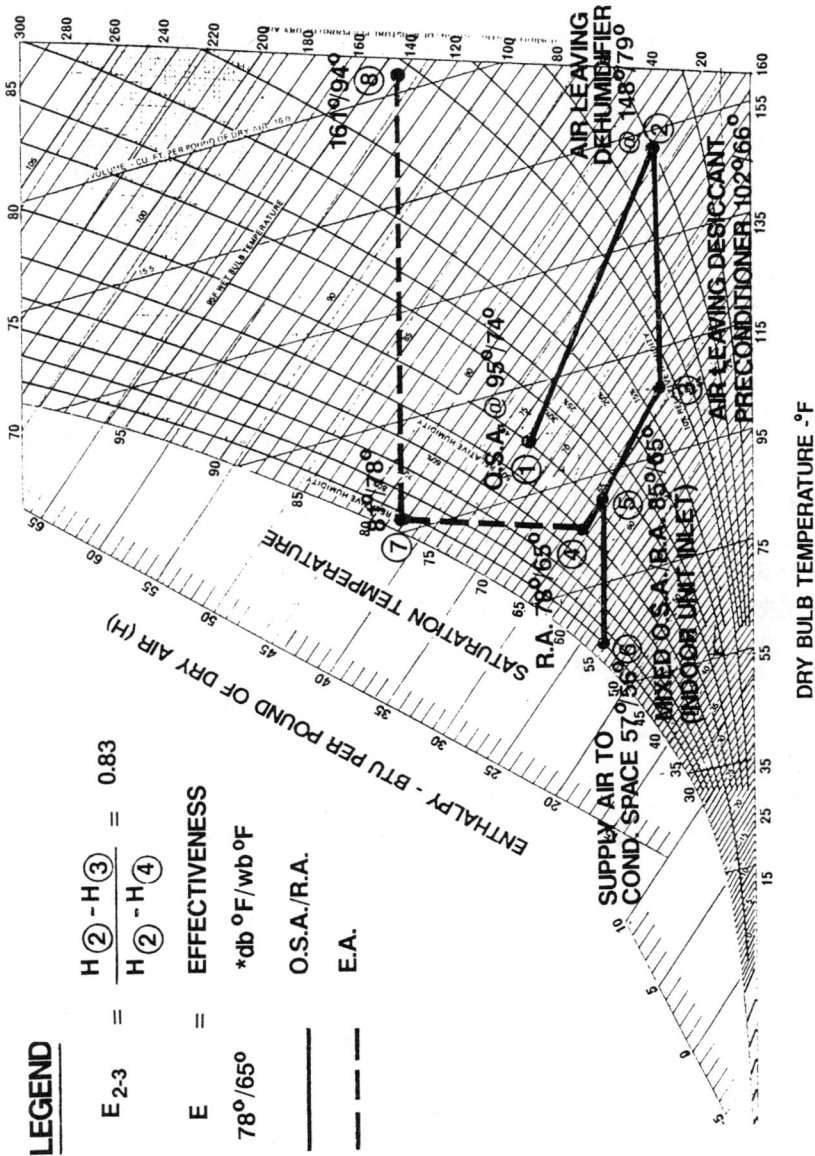

Figure 11-12. Psychromatic Chart for Desiccant-Assisted Distributed Rooftop Air-Conditioning System.

heating to the exhaust air prior to its entering the desiccant wheel regenerator section by piping interconnections with one of the adjacent rooftop condensing units in the manner shown in Figure 11-13. Figure 11-13 schematically shows the detailed refrigeration piping for the DDRA/C system.

Incorporating the evaporatively cooled heat-pipe module with evaporative cooling shown in Figure 11-3 into the desiccant-preconditioning module (in Figure 11-9), positions the desiccant wheel assembly to remove moisture from incoming outside air at 90 grains/lb bone-dry air (BDA) and deliver it at approximately 36 grains/lb BDA but with air leaving the desiccant wheel at approximately 148°F. The 53°F temperature-rise through the desiccant wheel represents the conversion of latent heat to sensible heat as well as thermal conduction from hot (regenerated) exhaust air passing through the regeneration wheel segment as shown in Figure 11-12.

The major operating cost of a desiccant dehumidification wheel is the reactivation cost, since moisture absorbed by the desiccant must be removed to reactivate the desiccant for subsequent reuse. In the DDRA/C system, efficiency at design conditions is improved by using the waste heat from the refrigeration condenser to heat the reactivation air. For units operating year-around, reactivation heat recovery is particularly effective during winter when reactivation inlet air temperatures are relatively low.

In practice, it is possible to provide heated regenerator air temperature up to 210°F with a positive displacement or scroll type displacement compressor employing the refrigerant R-22 (refer to Figure 11-7). The moisture removal capacity of the reactivation air depends on the prevailing outdoor weather conditions. However, for the required temperature of the reactivation for silica gel or LiCl, air is not likely to exceed 210°F for most comfort applications. Generally, only 50% of the energy recovered is required to completely reactivate the desiccant using preheated air at 190°F.

Modulating energy in response to actual latent loads is essential for cost-effective operations. Conventional desiccant systems are generally designed with adequate capacity to satisfy maximum loads. However, such maximum (latent) loads occur only for short periods depending on the microclimates and internal loads. Unless the system is equipped with modulating controls, it will consume the same amount of energy regardless of the actual combined external and internal load variations (refer to Figure 11-14).

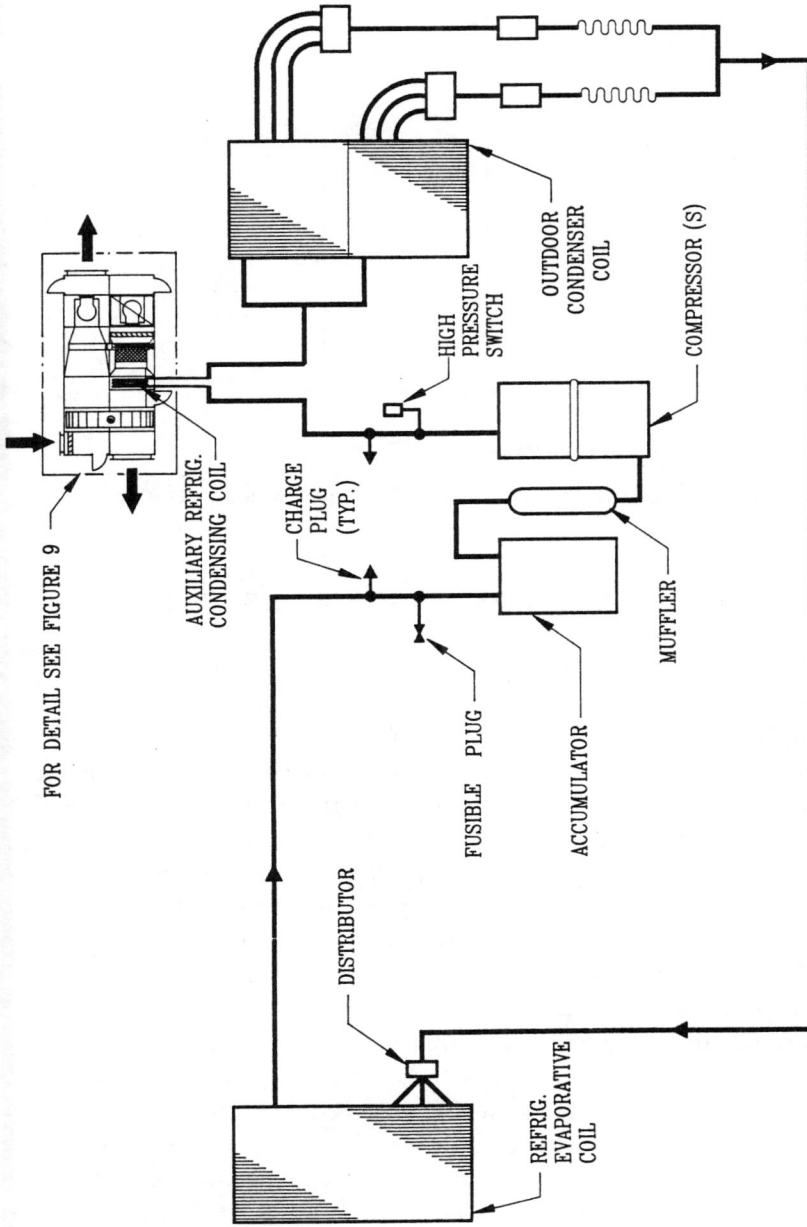

Figure 11-13. Refrigeration Piping for Desiccant-Assisted Distributed Rooftop Air-Conditioning System.

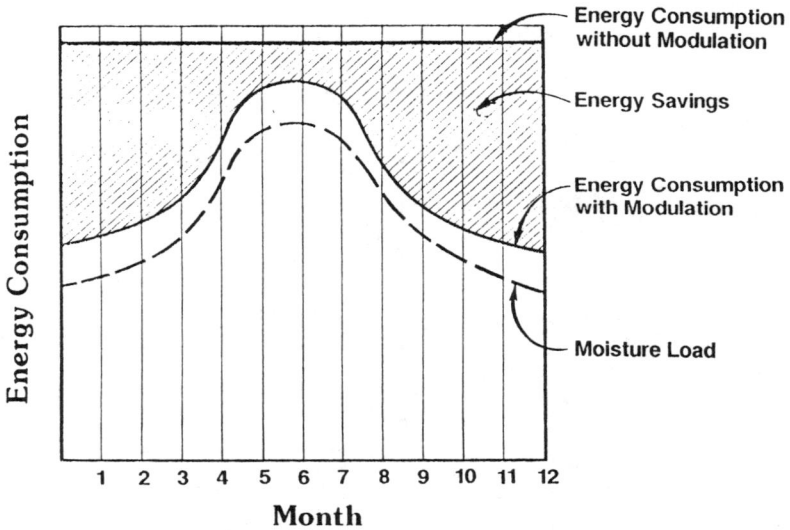

Figure 11-14. Effect of Modulation on Energy Consumption.

11.5 A PSYCHROMETRIC ANALYSIS FOR DDRA/C SYSTEM

Precooling hot outside air by means of the evaporatively cooled heat-pipe module shown in Figure 11-3 (and incorporated where shown in Figure 11-9), cools the discharge dehumidified outside airstream (Point 2 in Figure 11-9) to approximately 102°F upon exiting the desiccant-preconditioning module (Point 3) as shown in Figure 11-12. After mixing with space air (Point 4) at each DDRA/C indoor fan unit (Point 5), it is sensibly cooled to Point 6 before being delivered to the conditioned space. Notice that exhaust air leaving the heat-pipe spray section (Point 7) is then preheated by an auxiliary refrigerant coil and heated by an auxiliary gas (or electric) heater (Point 8) prior to entering the regenerated segment of the desiccant wheel for removal of absorbed moisture and its subsequent removal outdoors in exiting exhaust air (refer to Figure 11-9). In this way, the substantial total load associated with conditioning outdoor air described earlier (in Figure 11-1) is shifted from an energy-intensive vapor compression cycle serving downstream DDRA/C units to heat exhaust air for desiccant regeneration at significantly lower initial and operating costs. For comparison purposes, Figure 11-15 shows a psychrometric chart for the conventional RPA/C system.

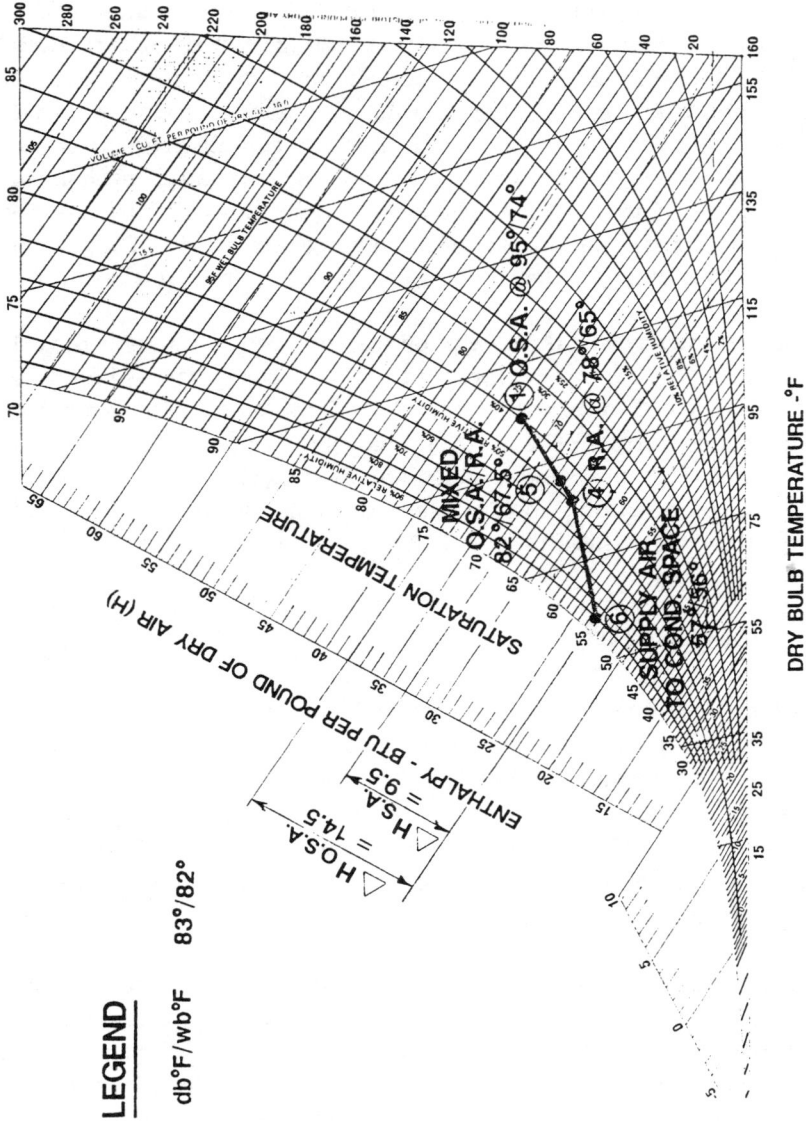

Figure 11-15. Psychrometric Chart for Conventional Rooftop Package Air-Conditioning System.

11.6 A COMPARATIVE ANALYSIS OF
DDRA/C AND RPA/C SYSTEMS

A comparative analysis between DDRA/C and conventional RPA/C systems for a prototype, one-story office building (11,000 ft² nominal) in five U.S. cities representing different microclimates was performed. These U.S. cities are Washington, D.C.; Atlanta; New Orleans; Chicago; and New York. Table 11-2 shows the total heating and cooling loads based on summer and winter design conditions, and the associated mechanical refrigeration load. In accordance with ASHRAE Standard 62-1989, a minimum outside air ventilation rate of 20 cfm/person for an office occupancy, and an occupancy distribution of 7 people per 1000 ft² were assumed.

The comparative analysis was performed using a state-of-the-art building simulation computer program. Also a comparative life-cycle analysis of DDRA/C and conventional RPA/C systems for the same five U.S. cities was performed. The life-cycle analysis assumed a 10-year useful life-span. Table 11-3 shows the results of this analysis. It takes into account the capital cost, operating and maintenance costs for each system, and shows the life-cycle cost differential between the two systems.

11.7 CONCLUSIONS

The comparative analysis performed has indicated an approximately 30%-savings in installed (or mechanical) refrigeration capacity regardless of the location in the U.S. This indicates that the installed refrigeration capacity of each conventional A/C unit served from a common desiccant-preconditioning module can be substantially downsized. Table 11-2 shows that the total heating load and total cooling air supply for DDRA/C system are less than the RPA/C system for all five U.S. cities.

As can be seen from Table 11-3, although the initial cost of the DDRA/C system is slightly higher than a comparable RPA/C system, the difference can be amortized in approximately a three-year payback period as the annual operating & maintenance (O&M) cost savings is substantially higher. The O&M cost savings ranges from $3,000 to $4,000 on a per-year basis for the five U.S. cities studied. This comparison is believed to be conservative since no credit has been taken for savings due to operation of

Table 11-2. Comparative Load Analysis for Conventional and Desiccant-Assisted Distributed Rooftop Air-Conditioning Systems for Offices in Five U.S. Cities.

	(1) SUMMER DESIGN CONDITION (db/wb,°F)	(2) WINTER DESIGN CONDITION (°F)	(3) MINIMUM OUTSIDE AIR (CFM)	TOTAL FLOOR AREA (FT^2)	TOTAL HEATING LOAD (BTU/HR)		TOTAL COOLING AIR SUPPLY (CFM)		MECHANICAL REFRIGERATION LOAD (TON)		(4) DPM COOLING LOAD (TON)
					(5) DDRA/C	(6) RPA/C	(5) DDRA/C	(6) RPA/C	(5) DDRA/C	(6) RPA/C	
ATLANTA	94/77	17	1,513	10,816	178,350	209,824	11,562	12,536	25.2	33.7	8.5
CHICAGO	91/77	−8	1,513	10,816	264,018	310,610	10,516	11,184	24.0	34.9	10.9
NEW ORLEANS	93/81	29	1,513	10,816	140,272	165,026	10,506	11,342	23.7	35.6	11.9
NEW YORK	90/76	12	1,513	10,816	198,353	233,356	10,393	10,951	23.3	31.2	7.9
WASHINGTON D.C.	95/78	14	1,513	10,816	200,533	235,921	10,491	11,586	23.8	34.2	10.4

(1) BASED ON 1% DRY–BULB AND 1% WET–BULB TEMPERATURES FOR SUMMER DESIGN CONDITIONS.

(2) BASED ON 1% DRY–BULB TEMPERATURE FOR WINTER DESIGN CONDITIONS.

(3) BASED ON 20 CFM/PERSON FOR OUTSIDE AIR QUANTITY.

(4) DPM   :   DESICCANT–PRECONDITIONING MODULE.

(5) DDRA/C   :   DESICCANT–ASSISTED DISTRIBUTED ROOFTOP AIR–CONDITIONING SYSTEM.

(6) RPA/C   :   CONVENTIONAL ROOFTOP PACKAGE AIR–CONDITIONING SYSTEM.

Table 11-3. Comparison of Life-Cycle Analysis of Conventional and Desiccant-Assisted Distributed Rooftop Air-Conditioning Systems for Offices in Five U.S. Cities.

	OPERATING AND MAINTENANCE COSTS ($)				CAPITAL COST ($)		CAPITAL COST DIFF.	TOTAL LIFE CYCLE COST ($)		LIFE-CYCLE COST DIFF.	INTERNAL RATE OF RETURN (%)
	FIRST YEAR		LAST YEAR								
	(2) DDRA/C	(3) RPA/C	(2) DDRA/C	(3) RPA/C	(2) DDRA/C	(3) RPA/C		(2) DDRA/C	(3) RPA/C		
ATLANTA	31,059	33,953	40,525	40,301	58,175 (4)	53,109	5,066	200,722	210,221	9,499	43.8
CHICAGO	15,766	18,324	20,570	23,909	58,175 (4)	53,109	5,066	127,029	134,913	7,884	38.6
NEW ORLEANS	23,520	25,965	30,688	33,505	58,175 (4)	53,109	5,066	164,395	171,733	7,338	36.6
NEW YORK	37,870	40,994	49,411	53,488	58,175 (4)	53,109	5,066	233,542	244,150	10,608	47.1
WASHINGTON D.C.	30,598	33,942	39,924	44,295	58,175 (4)	53,109	5,066	198,503	210,197	11,695	50.4

(1)   3% INFLATION RATE FOR OPERATING AND MAINTENANCE COSTS.

(2)   DDRA/C   :   DESICCANT-ASSISTED DISTRIBUTED ROOFTOP AIR-CONDITIONING SYSTEM.

(3)   RPA/C   :   CONVENTIONAL ROOFTOP PACKAGE AIR-CONDITIONING SYSTEM.

(4)   COST OF DESICCANT-PRECONDITIONING MODULE IS $7,500.

the heat-pipe assembly (located within the desiccant-preconditioning module) during the winter, which can be considerable in colder climates such as Chicago and New York. With RPA/C systems employing comparatively lower cost package components, when equipped with desiccant-preconditioning modules, further substantial savings may be realized for reasons given above. Table 11-3 also shows the life-cycle cost savings ranging from approximately $7,400 for New Orleans to approximately $11,700 for Washington, D.C.

In summary, the proposed DDRA/C system has the advantage of downsizing RPA/C evaporator coil and associated condensing unit capacities for comparable design loads, providing significant annual O&M cost savings, and cooling with the use of reduced chlorofluorocarbons (CFCs). The DDRA/C system allows independent temperature and humidity controls and, therefore improved IAQ and enhanced occupant comfort.

11.8 REFERENCES

[1]Marciniak, T. et al., "Gas-Fired Desiccant Dehumidification System in a Quick-Service Restaurant," *ASHRAE Transactions*, Vol. 97, Pt. 1, Heating, Ventilating and Air-Conditioning Engineers, Inc., Atlanta, GA, 1991.

[2]Busweiler, U., "Air Conditioning with a Combination of Radiant Cooling, Displacement, Ventilation, and Desiccant Cooling," *ASHRAE Transactions*, Vol. 99, Pt. 4, Heating, Ventilating and Air-Conditioning Engineers, Inc., Atlanta, GA, 1993.

[3]Belding, W. et al., "Desiccant Development for Gas-Fired Desiccant Cooling Systems," *ASHRAE Transactions*, Vol. 97, Pt. 1, Heating, Ventilating and Air-Conditioning Engineers, Inc., Atlanta, GA, 1991.

[4]Pesaran, A. and R. Anderson, "Innovative Solid Desiccant Substrates for Desiccant Dehumidifiers," *ASHRAE Transactions*, Vol. 97, Pt. 1, Heating, Ventilating and Air-Conditioning Engineers, Inc., Atlanta, GA, 1991.

[5]Matsuki, K. and Y. Saito, "Desiccant Cooling R&D in Japan," *ASHRAE Transactions*, Vol. 94, Pt. 1, Heating, Ventilating and Air-Conditioning Engineers, Inc., Atlanta, GA, 1988.

[6]Pesaran, A. and T. Penny, "Impact of Desiccant Degradation on Desiccant Cooling System Performance," *ASHRAE Transactions*, Vol. 97, Pt. 1,

Heating, Ventilating and Air-Conditioning Engineers, Inc., Atlanta, GA, 1991.

[7]Meckler, M., Desiccant Assisted Air Conditioner, U.S. Patent 4,887,438, 1989.

[8]Patents pending.

[9]Meckler, M., "Integrated Desiccant Cold Air Distribution System," *ASHRAE Transactions*, Vol. 95, Pt. 2, Heating, Ventilating and Air-Conditioning Engineers, Inc., Atlanta, GA, 1989.

[10]Peterson, J., "An Effectiveness Model for Indirect Evaporative Coolers," *ASHRAE Transactions*, Vol. 99, Pt. 2, Heating, Ventilating and Air-Conditioning Engineers, Inc., Atlanta, GA, 1993.

Section 4

Operation and
Maintenance Procedures
To Improve
Indoor Air Quality

12

Maintaining Acceptable IAQ During Renovations

Francis M. Gallo, P.E., CFM, CIAQP
Vice President, LZA Technology
New York, New York

12.1 INTRODUCTION

Renovations in the workplace are potentially the greatest source of indoor air quality (IAQ) problems a building may face. It is during renovations that construction dust is created and dispersed, and when paints, adhesives and solvents are introduced and will typically offgas at maximum rates. Building renovation is also a time when existing ceiling grids and heating, ventilating and air-conditioning (HVAC) systems are disturbed either by being replaced or modified, causing years of accumulated dirt, some of which may contain microbial contamination, to become airborne. Additionally, it is a time when the number of people involved in renovation activities can tax management strategies that are normally adequate to control IAQ during routine occupancy periods. Therefore, the challenge facing the building and project manager is to understand all the forces that can affect IAQ during renovations and develop strategies to deal with them.

12.2 RENOVATION INTERESTS AND CONSTRAINTS

Many interests and constraints exist during the planning and construction phases of a renovation (refer to Figure 12-1). The interests of building owners, facility managers, contractors, technicians, government agencies and tenants can affect the way renovations are planned and performed. Capital, building and HVAC design, occupancy, and space layout are factors and constraints that may limit the application of technology to maintain acceptable IAQ. All of these factors must be understood and their influence anticipated by the building and project managers for a project to be successful. Technological strategies must be selected to deal with the physical constraints and management strategies must be selected to deal with the varied interests of the parties involved.

12.2.1 Technological Strategy

Technological strategies to maintain acceptable IAQ during renovations should consist of at least three fundamental components: (a) isolating the renovation area from occupied areas, (b) minimizing contaminant generation (including dust, volatile organic compounds [VOCs] and noise), and (c) maintaining building and HVAC system hygiene.

Figure 12-1. Factors and Constraints Affecting Application of Technology to Maintain Acceptable Indoor Air Quality.

12.2.1.1 Isolation

Isolation of the work area is the first line of defense for maintaining acceptable IAQ in occupied areas. Work should preferably be performed during off-hours when renovating near or in occupied areas. If possible, the work area should be under "negative pressure" relative to the surrounding occupied areas not under renovation. Possible values for the relative negative pressure depend on the type and magnitude of contaminant generation and may range from what is normally recommended for normal building pressurization to the external environment (on the order of 0.05 inch of water) to much larger values as required on asbestos remediation projects.

Negative pressurization may be achieved by sealing off the construction zone and implementing local exhaust directly to the outside via a knock out in a window or wall. When local exhaust is not possible, high-efficiency particulate air (HEPA) filters and, if necessary, gaseous filtration may be an alternative. In these cases, the return air from the construction area is filtered to prevent construction dust and other particulate matter from entering the HVAC system and being dispersed throughout the building. However, effective gas-phase filtration is often impractical or prohibitively expensive. In this case, selection of low VOC-emitting materials and scheduling of construction work during unoccupied periods may be the only practical alternative. It is also important to protect construction workers. In addition to the Occupational Safety and Health Administration (OSHA) requirements for construction safety, manufacturers recommendations for construction materials should be closely followed.

12.2.1.2 Minimizing Contaminant Generation

The selection of construction materials, procedures and methods can have a dramatic effect on the concentration of indoor air contaminants generated during a renovation. It is much better for both construction workers and building occupants to reduce contaminant generation than to either exhaust or filter it. Proactive material selection is a prudent first step. The designer should specify office products, construction materials, furnishings, paints, solvents and adhesives with low-emission characteristics. Factory or warehouse offgasing of office furnishings and construction materials prior to installation is another contaminant source management technique. One can resort to several resources for specifying materials of construction and understanding their effect on IAQ. For example,

The Environmental Resource Guide, published by American Institute of Architects is an excellent resource. Construction methods can also reduce contaminant levels. For example, when saw-cutting concrete, wetting of the concrete with water prior to cutting can reduce the amount of dust generated.

12.2.1.3 Maintaining Building and HVAC Hygiene

It is important to maintain the building and HVAC system hygiene both during and immediately after a renovation. Construction dust and other contaminants can find their way through unknown or unanticipated pathways even when isolation of the construction area is thought to be complete. It may be prudent to upgrade the HVAC system filtration during a renovation and/or place local filtration on return grills near the construction area. During construction, the renovation area and its immediate surroundings should be HEPA vacuumed periodically to minimize the migration of construction dust into occupied areas in the HVAC system. If possible, the HVAC system serving the construction area should be shut down and isolated, and an alternate means of ventilating area be employed (e.g., by using local exhaust and transfer air from an adjacent area). One must make sure that the commissioning procedures recommended by ASHRAE Guideline 1-1989 before, during and after the renovation are properly employed. This will ensure that the renovated systems operate in accordance with the design intent, and that the system and area are properly cleaned and decontaminated prior to occupancy.

12.2.2 Management Strategy

The management strategy employed to deal with the varied and, at times, competing interests of the parties involved in the renovation process is just as important as the technological strategy. The management process consists of four elements: (a) organizing, (b) planning, (c) controlling, and (d) leading (refer to Figure 12-2). The building or project manager should tailor these elements to fit the project goals.

12.2.2.1 Organizing

IAQ responsibilities should be organized and assigned to individuals or groups such as the building manager, project manager, construction supervisor, tenant representative, etc. The assignment of responsibilities must be carefully linked into a team effort with IAQ responsibilities carefully defined. Each project team member must understand his or her IAQ responsibilities and those of other team members. Please refer to

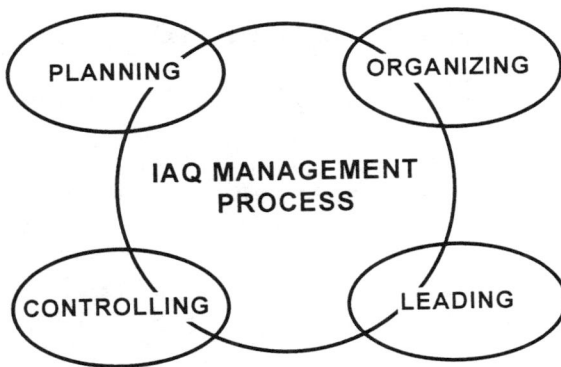

Figure 12-2. Elements of Indoor Air Quality Management Process.

Table 12-1 for a sample IAQ responsibility chart. For example, suppose that a parking garage in the northeast requires the installation of an elastomeric coating to prevent the intrusion of road salts into the concrete slab, which was causing spalling and will cause in structural failure of the parking deck, if not stopped. The selected coating contains a volatile compound that off-gasses during the first six hours of the curing process. This compound can produce respiratory distress in sensitized individuals. Alternate coating systems without the compound of concern were evaluated and rejected because these alternate coating systems did not possess the required elastomeric characteristics. The garage is adjacent to the air intakes of its associated office building. The project manager decides to institute an IAQ plan to protect building occupants.

The building manager, in consultation with the project team, decides that the work should take place at night since there was no effective way to eliminate the introduction of the volatile compounds into the building via the air intakes and building infiltration. According to a procedure instituted, building occupants are asked to leave prior to start of the work each night. External entrances of the building will be sealed using polyethylene and duct tape, and the HVAC systems will be shut down. The contractor is responsible for ensuring that his workers have personal protective equipment and that each night work will not start prior to the "all-clear" given by building night superintendent. The night superintendent's responsibilities include ensuring that the HVAC system was off, the doorways sealed, and that all occupants were evacuated from the building. He is also responsible for ensuring that the work stopped by midnight. The security manager's responsibilities include a sweep of the

Table 12-1. A Sample of Indoor Air Quality Responsibility Chart.

IAQ Team Member	IAQ Role	IAQ Responsibilities
Building Manager	Plans, Organizes, Leads, And Controls	• Ultimate Responsibility for IAQ • Has Approval And Veto Power of IAQ Strategies • Provides Resources Necessary to Perform Project According to IAQ Plan
Project Manager	Coordination of Project Activities And Resources	• Determine IAQ Strategy Efficacy • Inform Building Manager of IAQ Status • Ensure All IAQ Policies Are Followed • Complaint Resolution
Construction Supervisor	Coordination of Tasks	• Monitors Worker Compliance with IAQ Plan • Ensures All Safety Rules Are Followed • Reports Problems with IAQ Strategy to Project Manager
Tenant Representative	Provider of Feedback from Occupied Spaces	• Logs and Reports Any IAQ Complaints • Coordinates Access to Occupied Space
IAQ Consultant / Health And Safety Manager	Provider of Expert IAQ Information	• Provides Input in Developing IAQ Plan • Recommends IAQ control methods • Monitors IAQ • Assists in Pre And Post Occupancy Commissioning

building to advise people that the work was beginning in an hour and that they should leave the building. The IAQ consultant's responsibilities include setting up real-time monitoring devices at strategic locations to monitor the compounds of concern.

It is important to note that IAQ obligations cannot be delegated. Only tasks can be delegated. The obligation to maintain acceptable IAQ during renovation remains with the building or project manager. In the above example the project manager had the ultimate responsibility to ensure that each team member followed his/her responsibilities.

12.2.2.2 Planning

Planning to maintain acceptable IAQ during renovations requires establishing IAQ policies, strategies and schedules. Policies are necessary to guide those involved in activities that affect IAQ. Policies provide the project team with a direction on how to conduct a renovation. The broadness of the policy will vary according to individual project circumstances. Some examples of IAQ policies are: *"All painting work performed in occupied areas shall be accomplished off-hours to minimize exposure to occupants," "Only low VOC-emitting materials shall be used in a renovation,"* and *"Measures shall be taken during renovations to minimize exposure to occupants of pollutants generated by the renovation process."* The first two example policies provide specific direction while the third example provides only general guidance.

Strategies are developed by predicting possible IAQ problems, finding possible solutions, evaluating each possible solution and selecting an appropriate course of action. Predicting potential IAQ problems requires understanding the sources of contaminants, their pathways into occupied areas, and their effect on building occupants. There may be several solutions to the problem ranging from selection of construction processes and materials that do not produce indoor contaminants of concern to evacuation of the building during renovation. In most cases, the solutions will encompass elements of proactive source reduction and pathway control.

Challenges in developing strategies include setting a satisfactory target level of IAQ achievement, narrowing the number of alternative solutions, and the method selected for decision making. The project manager should seek guidance, decide what level, if any, of contaminant generation and introduction into occupied areas is acceptable. Are there any legal, health, comfort or company criteria available? Do the permissible exposure limits for specific substances become the control or one-tenth of the threshold limit values? Are there strategies available to limit the exposure of generated contaminants as close to zero as possible? Narrowing the number of solutions will depend on their feasibility, timeliness and economy. Lastly, the steps taken for decision making should be based on a process that anticipates possible problems and allows for mid-course corrections (refer to Figure 12-3).

Schedules are developed and implemented to ensure that activities are performed in the proper sequence and that adverse IAQ consequences are minimized. Careful scheduling of renovation work can be most effective in maintaining acceptable IAQ. Deferring those activities that have

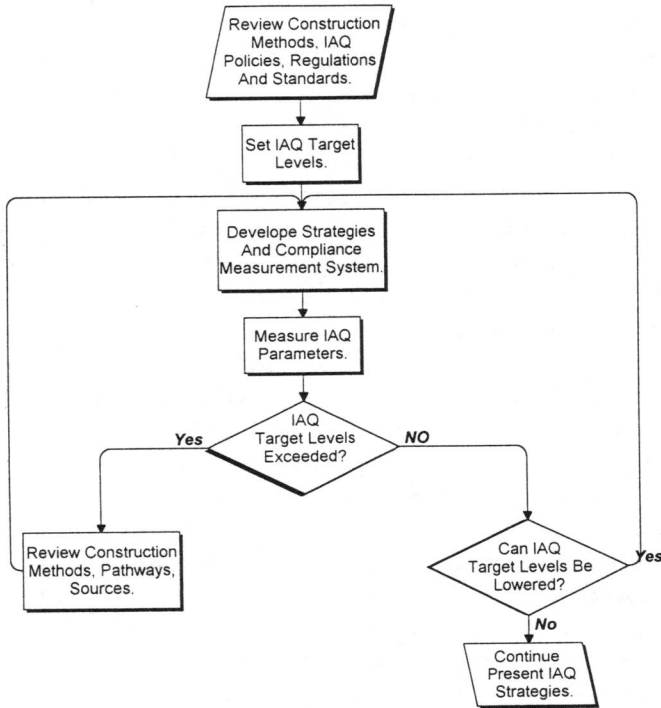

Figure 12-3. A Decision Process for a Renovation Project.

the potential to produce the most contaminants to unoccupied periods may be more effective than other measures and ultimately more cost-effective.

12.2.2.3 Controlling

Renovation activities must be controlled to achieve the desired IAQ goals. This requires obtaining timely information to help assess the efficacy of the proactive IAQ measures taken. There are three basic forms of control: steering controls, yes/no checks, and post-action evaluations. Steering controls correct procedures during the renovation by comparing actual outcomes with predicted outcomes. Variances are analyzed and corrections are made to the renovation process. For example, an IAQ plan may assume that placing an area under renovation using local exhaust will control contaminant migration into occupied areas. This decision should be tested during the actual renovation. If it is found that contaminants are finding their way into occupied areas, one should have an alternative solution ready such as using different construction techniques or rescheduling work during unoccupied hours.

Steering controls provide the best form of control because they can measure the efficacy of the control measures as the work progresses in an area. Unfortunately, this may not be possible in all circumstances. Certain contaminants can only be measured by sending samples to a laboratory. By the time laboratory results are in, the project may have progressed to the next phase or area. This leads us to the next type of control, yes/no checks. Yes/no checks again involve comparing the expected result with the actual outcome. However, because of the time lag in measuring contaminant levels, a determination of the efficacy of the control measures is made after a portion of the work is completed but prior to moving on to the next area. Yes/no checks are well suited to renovation activities that are quite repetitious, such as renovating a suite of offices in sequence. Lessons learned from the first office can be applied to subsequent ones.

The third form of control closely related to yes/no checks, is the post-action evaluations. The evaluation of the efficacy of IAQ measures is made only after the project is completed. This control system is normally the least effective but may be the only practical type when dealing with single "quick hit" type projects such as emergency roof repairs, carpeting a room, etc. Lessons learned by this method are applied to similar future renovations.

Regardless of what control scheme is followed, the project manager must decide who will control and when, and what measurement techniques should be employed to verify compliance. IAQ parameters that are easily and commonly measured such as carbon dioxide concentrations may be an improper choice during renovations. There may be a temptation to use occupant complaints (or the lack of them) as a control system. This approach should be avoided. First, it may jeopardize health. Secondly, the occupant complaints may not reach the project team in a timely manner. By the time the project team reacts, there may be a serious IAQ problem.

12.2.2.4 Leading

The leadership exhibited by the project manager is key to the successful renovation. He/she must predict the responses of individual team members, overcome any opposition and change the IAQ strategy as conditions warrant. Communication skills, motivation techniques and the effective use of compromise and coercion are required to successfully oversee a renovation. In short, leadership involves putting planning, organizing and controlling into action and, "making it happen."

There are three elements in providing leadership in providing acceptable IAQ during renovations. The project manager must predict the response of individual team members to the IAQ strategy. Each party involved in the renovation process will have their own interests and needs at stake. What is important to one member of the team may not be of any consequence to the others and, of course, maintaining acceptable IAQ during renovation may not immediately enhance profit margins. The contractor may have bid the job on slim margins and may look for short-cuts to complete the job. The tenant may be under time constraints to complete a particular renovation by a certain date. Specified materials of construction that were specifically selected for their low effect on IAQ may not be in stock and substitutes may need to be used. Therefore, understanding needs and interests and predicting responses will help in overcoming opposition and indifference to the IAQ plan.

An appropriate motivating strategy should be selected to overcome anticipated opposition to the IAQ plan. Incentives may be necessary to get the project team to act in the desired manner. Such incentives can be monetary or non-monetary. For example, the promise of future work may often convince the contractor that it is in his best interest to follow the IAQ plan. Conversely, strict monitoring by the project manager with the threat of termination of contract may be necessary in other cases. In any event, project manager must exhibit the leadership and fortitude to deal with probable resistance.

Finally, strategies may have to be changed when actual responses differ from anticipated responses. The threat of termination of contract may not produce the desired result if the contractor has a backlog of work and no intention of performing work in the building again. In those cases, compromise may be the only viable strategy. In conclusion, maintaining acceptable IAQ during renovations is as much a management challenge as a technical one. Traditional management principles, sound technical controls, effective communication skills and leadership will often be necessary to get the job done.

12.3 REFERENCES

[1]Bearg, D.W., *Indoor Air Quality and HVAC Systems*, Lewis Publishers, p. 17.
[2]Newman, W., *The Process of Management*, 4th ed., Prentice-Hall, Inc.

13

Proactive IAQ Building Management

Francis M. Gallo, P.E., CFM, CIAQP
Vice President, LZA Technology
New York, New York

13.1 INTRODUCTION

Indoor air quality (IAQ) has become an issue of concern in recent years for employers, building owners and managers, unions, and tenants. The Environmental Protection Agency (EPA) considers inadequate IAQ a significant risk. Recently, the Occupational Safety and Health Administration (OSHA) has proposed sweeping new rules that will require employers in the U.S. to develop and implement IAQ programs in non-industrial workplaces. While some states are considering legislation regulating IAQ, a few states such as New Jersey have already passed legislation. Moreover, IAQ litigation has become more frequent in recent years. There are several reasons why IAQ has become a great concern and problem. Some of these reasons are discussed below.

People today spend up to 90% of their lives indoors. Very little time is spent outdoors in a typical day. Also, there is a greater variety of contaminants found indoors than outdoors, and indoor air contaminant concentrations are typically higher than those outdoors. Thus, indoor air contaminants have a greater probability of affecting most people.

The indoor environment has changed. Unwanted infiltration and design ventilation had been reduced in the 1980s to save energy. Consequently, the amount of outside air entering a building can be significantly less than necessary to dilute contaminants to desired concentrations. Only in recent years have building codes been revised to mandate an increase in the amount of ventilation air introduced into a building. Additionally, there is a multitude of cleaning products, maintenance materials, office machines, furnishings and tenant activities that can emit particulate and gaseous contaminants. Copy machines, laser printers, correction fluids, floor waxstrippers, furniture restoration chemicals, pesticides and environmental tobacco smoke (ETS) are examples of contaminant sources found in a typical office today. Also, furnishings can become "sinks" for particulate and gases absorbing and releasing contaminants continually.

Building systems can affect IAQ to a great degree. These systems are the heating, ventilating and air-conditioning (HVAC) systems; potable water system; and the energy management, etc. These systems must be properly designed, operated and maintained. Otherwise, they can become sources of IAQ problems. Unfortunately, decreasing maintenance budgets and personnel and increasing energy costs can result in inadequate resources allocated to maintaining building systems. Additionally, most maintenance procedures and recommendations are specified to maintain design service life and not necessarily to maintain acceptable IAQ. For example, filters that are adequate to protect the cooling coils from gross fouling may not be satisfactory for IAQ purposes.

Employers' and employees' expectations for their work environment have changed. Most office employees will not tolerate an uncomfortable or unhealthy working environment. Conditions that affect employee productivity will quickly catch management's attention. Causes and solutions will be sought. There is also a broad cross section of people in various degrees of health. It is not unusual to find people of various ages with allergies, asthma, heart conditions and other ailments. These people may react to low concentrations of indoor air contaminants and to a greater degree than healthy young adults.

IAQ litigation in recent years has become more common and worrisome to not only building owners but property management firms, design firms, contractors, maintenance and cleaning companies, and manufacturers. A landmark liability case was litigated in 1990. In the Call vs. Prudential case, two tenants in a building alleged loss of employee productivity and business losses due to renovations on another floor. The judge ruled that the building and its systems (e.g., HVAC) could be considered products and thus the manufacturer(s) of such products could be held liable. Thus, the designers, engineers, construction contractors, equipment manufacturers, building operators and others were subject to liability. This case was settled for an undisclosed sum.

IAQ legislation, regulation and standards of care are being promulgated. The most recent example of this is the IAQ rule proposed by OSHA. This rule, if promulgated, will affect every commercial property owner and operator in the U.S. The rule will require employers to establish documented IAQ plans for non-industrial buildings and assign a designated person to oversee the IAQ plan. In many cases, the building operator will be the designated person for all the tenants. This will place a tremendous responsibility on building owners and managers to be knowledgeable in IAQ and have proactive programs in place.

Another example of impending government involvement is the EPA's Building Air Quality Alliance Program. This is a voluntary program in which building owners subscribe to a set of guiding principles in IAQ and agree to operate their buildings to promote adequate IAQ. The four guiding principles are (a) making IAQ a priority, (b) knowing what to do to prevent indoor air contamination, (c) practice sound IAQ management, and (d) take care of problems that occur. The program requires basic compliance with the IAQ practices and procedures delineated in the EPA's Building Air Quality book[1].

The engineering community is also revising standards that will affect how buildings are designed and maintained such as ASHRAE Standard 62-1989: *Ventilation for Acceptable Indoor Air Quality*. This standard specifies the amount of outside air that must be brought into a space to dilute contaminants. New additions to the standard (presently being revised) will most likely specify specific maintenance practices to be followed.

Unions are very active on the IAQ issue. Unions that represent office employees, teachers, government employees, and trades people that operate and maintain buildings are pressuring government to enact IAQ regulations to protect their employees. IAQ will continue to be an important issue for building operators, owners and tenants for many years to come. The science of IAQ is relatively young. What is accepted as a good practice today may be an inadequate practice tomorrow. The building managers and engineers must keep up with latest developments in IAQ and develop proactive programs based on the latest findings, recommendations and requirements issued by IAQ researchers, experts and government bodies.

13.2 FACILITY FACTORS AND STRATEGIES

There are many facility-related factors that affect IAQ in a workplace. They can act singly or simultaneously to adversely affect or favorably enhance the indoor environment. The building manager should develop a list of possible factors for each building. Understanding these factors is key to developing a successful proactive program. These factors include HVAC design, maintenance and hygiene, office cleaning practices, contaminant source management, energy management, space layout, tenant communications, renovation practices, proactive inspections, and IAQ training for management and operating staff.

Proactive strategies should be developed to address these factors for each building. The strategies should be reviewed periodically and updated to incorporate government regulations and the latest scientific findings of IAQ researchers and practitioners.

13.2.1 HVAC System

The HVAC system of a building is a key factor. HVAC system design and how it is operated and maintained can affect IAQ to a greater degree. Improper commissioning, insufficient design, and years of inadequate maintenance can subvert a building's ability to provide adequate IAQ. Upon investigation, a building manager may find that some diffusers have never been connected, outside air dampers are inoperative, air-handling unit (AHU) coils are dirty and undersized, ductwork is contaminated, and filtration is inadequate.

The building manager should determine which ventilation standard the system was designed to comply with and, more importantly, what the present ventilation and thermal performance capabilities of the system are. Ideally, the HVAC system should be able to provide ventilation in accordance with ASHRAE Standard 62-1989. Of course, the building code may not require a building owner to upgrade the system in accordance with the latest ventilation standard unless a renovation takes place. However, upgrading in accordance with the latest standard may be a prudent measure regardless of the lack of any legal requirement. An engineering study may be required to determine if compliance with ASHRAE Standard 62-1989 is feasible with the existing system. If not, energy saving strategies such as heat recovery equipment and desiccants should be explored to offset upgrading costs. HVAC controls should be calibrated and the entire system should be balanced if necessary. The hygiene of AHUs and ductwork should be determined and improved if warranted.

13.2.2 Cleaning Practices

Another facility-related factor is that how often and how well the occupied space is cleaned. While almost all floors and horizontal work surfaces are cleaned regularly, vertical and upholstered surfaces may receive inadequate treatment. Vertical and upholstered surfaces can accumulate particulate matter and allow their release when disturbed or touched. Additionally, the cleaning staff may be equipped with low-efficiency vacuum cleaners that do not capture particulate matter adequately in the respirable range (below 5 microns).

Conventional vacuum cleaner efficiency ranges between 30% and 60% for particle sizes of 1 micron and larger. Ordinary vacuum cleaners experience significant leakage of respirable sized particles. Higher efficiency vacuum cleaners and/or vacuum bags now coming onto the market can dramatically increase collection efficiency to over 98% and may offset a reduced cleaning schedule forced by budget constraints in many companies. Vertical and fleecy surfaces should be cleaned. Manufacturers' recommendations concerning ventilation requirements of cleaning compounds should be strictly followed.

13.2.3 Contaminant Source Management

Contaminant sources and their management is another factor that the building manager should examine. Some office equipment may offgas chemicals or particulates. Materials used in construction and maintenance may require special consideration during application or use. Occupants themselves can intentionally or unwittingly introduce contaminant sources into the space that may affect overall air quality and tax the capabilities of the HVAC system to remove them.

Processes or equipment that introduce contaminants may require isolation from the main HVAC system. In those cases, a separate, dedicated exhaust may be required. Methods for addressing occupant-introduced contaminants should be explored including development of building policies, occupant education, changes in cleaning methods and frequencies, and modification of the HVAC system to reduce possible harmful concentrations of such contaminants. The ventilation requirements and precautions listed in the Material Safety Data Sheets (MSDS) of chemicals used should be strictly followed.

13.2.4 Space Layout

How the occupied space is laid out and used can affect IAQ. Improperly placed office furniture and partitions can adversely affect the ventilation effectiveness in the space. Overcrowding can burden the HVAC system in providing adequate ventilation. Revisions in space layout without considering the HVAC system can leave offices without supply and/ or return.

Before undertaking any space changes, the building manager and space planner should understand the assumptions made by the original architects and engineers for space use. This will give valuable information on the capabilities of the existing HVAC system to handle contemplated space changes. Once the design assumptions are understood, space changes should be planned so that ventilation effectiveness, and outside airflow rate for the space is adequate. In some cases, provisions for supplemental ventilation may be required (e.g., converting a group of single offices to a large conference room).

13.2.5 Tenant Communications

How the building manager communicates with building occupants and the methods of communication employed can determine if an IAQ concern turns into a crisis. Lack of adequate communication between management and employees or tenants can leave the impression of an uncaring or ignorant building staff prolonging and intensifying anxiety about IAQ.

Several avenues of communication should be set up to advise tenants of the building IAQ program. Written forms of communication may include publishing details of the IAQ program in a building newsletter, postings on bulletin boards, or special employee communications. Oral communications may consist of IAQ presentations at safety meetings, special presentations where all departments or floors are invited for one-on-one consultations. During an IAQ investigation, it is important not to underreact or overreact. A reliable IAQ communication plan fashioned after other corporate communication plans should be enacted.

13.2.6 Renovations

Renovations have the potential of being the most serious IAQ factor. During renovations, construction dust is released and dispersed, and paints, adhesives, and solvents are introduced, some of which may offgas during curing. It is also a time when ceiling grids and HVAC systems are disturbed by being replaced or modified, possibly causing years of accumulated dirt, some of which may contain microbial contamination, to become airborne. Additionally, it is a time when the number of people involved in a renovation can tax management strategies that are normally adequate to control IAQ during routine occupancy.

Before renovation, possible sources of contaminants and their pathways into occupied spaces should be explored. The renovation area should be isolated from occupied areas, methods to reduce contaminant generation (including noise) should be employed, and HVAC system hygiene should be maintained. Ways to isolate the renovation area from occupied areas include negative pressurization, local exhaust, and high efficiency particulate and gaseous filtration. Another strategy involves performing work during unoccupied times. However, care should be exercised to prevent contaminants from surviving until the next occupied period.

The building manager should ensure that all parties involved in the renovation process, including contractors, tenants, maintenance personnel, and safety compliance officers, have a clear delineation of IAQ responsibilities. Obligation to ensure acceptable IAQ remains with the building manager and cannot be delegated. Only specific IAQ tasks and preventive measures can be delegated.

13.2.7 Proactive Inspections

The condition of the facility and its HVAC system need to be continuously assessed. IAQ strategies that rely solely on responding to complaints without a proactive facility assessment may induce unnecessary complaints and costs. The building manager may find that qualitative and quantitative data regarding HVAC system operation & maintenance (O&M), HVAC system hygiene, and levels of key indicative contaminants provide invaluable feedback. Another added benefit of building system facility assessments is that conditions that may cause inadequate IAQ can also contribute to excessive energy consumption. Thus improving IAQ may also improve operating efficiency.

A proactive inspection program should be developed and implemented. This program should contain a statement of goals and list of parameters that will serve as measures of goal attainment. Inspections may include daily visual confirmation of AHU operation, filter fit, hygiene, monthly examinations of thermal complaint trends, and semi-annual examinations of ductwork. A more extensive and comprehensive inspection may include semi-annual measurements of key indicative parameters including, but not necessarily limited to, temperature, relative humidity and carbon dioxide.

13.2.8 IAQ Training

IAQ is a complex subject. All parties involved in managing and working in a facility should be familiar with conditions that can adversely affect IAQ. The building manager, health and safety manager, contract supervisors, O&M personnel and office occupants can benefit a great deal from IAQ training programs that are available from professional societies, trade associations, colleges and universities.

A plan must be developed to provide training for key staff and all operating personnel. All maintenance personnel should receive at least two hours of training while those responsible for administering the facilities IAQ program (e.g., building manager, health and safety manager, etc.) should receive at least six hours of training. The strategies outlined above can be combined to design a proactive IAQ program that is comprehensive and site-specific. However, the program should be reviewed continuously and updated to account for the latest scientific findings and government regulations. Continuous improvement in IAQ is possible with adequate resources, proactive maintenance, management commitment, employment of air-cleaning technologies, and training.

13.3 OPERATING HVAC SYSTEMS FOR IAQ

When operating an HVAC system, consideration should be given to health-related factors as well as comfort. Compliance with local building codes does not necessarily ensure that the system will minimize the potential for IAQ problems. Additionally, there are many factors not addressed by the applicable codes that will produce a more productive work environment.

Some of the HVAC system operating practices that may minimize the potential for HVAC system-related IAQ problems are discussed below. This discussion is intended to make building managers and operators aware of some important operational issues. It is not intended to be a comprehensive listing of all issues or to suggest specific operational procedures that can only be developed on a building-specific basis.

13.3.1 Compliance with ASHRAE Standard 62-1989

ASHRAE Standard 62-1989 specifies minimum ventilation rates that will provide acceptable IAQ to occupants and will minimize the potential for adverse health effects. For office occupancies, for example, the standard specifies a minimum of 20 cubic feet per minute (cfm) per person of outside airflow rate based on a maximum occupancy of 7 persons per 1000 ft^2. Ventilation effectiveness, the fraction of the outside air that reaches the occupied zone, in some cases, is less than 100% (short circuiting). In this case, the outside airflow rate required will be greater than 20 cfm. The building engineer should examine diffuser and office partitioning layouts for conditions that result in short circuiting.

Ventilation systems should also provide the requisite amount of outside air while the building is occupied. Many systems vary the amount of outside air introduced to the space and may underventilate the space under some operating conditions. Variable-air-volume (VAV) systems, by the nature of their design, may underventilate the space when thermal conditions are satisfied. Many VAV boxes will close down to very low- or no-flow conditions. The building engineer should examine any VAV system in the building and determine if the boxes close down completely. Merely installing minimum position stops may result in unacceptable temperature control. An engineering study may be necessary to properly design a retrofit that will not compromise temperature regulation.

13.3.2 Compliance with ASHRAE Standard 55-1992

ASHRAE Standard 55-1992 specifies thermal conditions that will satisfy most occupants of a space. Conditions considered by the standard include temperature, humidity, thermal radiation, and air speed. The building engineer should operate the HVAC system in a manner that provides compliance with this standard in all seasons of the year. Humidity regulation in the winter will be a problem in buildings without humidification systems. In those cases, the only acceptable alternative will be the installation of a well-designed humidification system. However, extreme care should be exercised in specifying humidification systems since in many cases inadequate maintenance and improper design can cause IAQ problems (refer to 13.3.4 Section).

13.3.3 Adequate Filtration

Filtration is very important in ensuring that the ventilation system remains clean and does not become a source of potential microbial contamination. Adequate filtration of particles in the range of 0.5 micron and smaller is very important. These respirable particles may contain fungal spores, bacteria and lung-damaging dust, and can be inhaled deeply into the lungs. Medium efficiency filters (40%, ASHRAE dustspot) should be used at a minimum. Eighty-five percent and higher efficient filters are highly desirable. However, any added pressure-drop of the higher efficiency filters should be determined for compatibility with the AHU. Also, many small heat pump, fan coil, and unit ventilator systems do not have filter channels that can accommodate higher efficiency filters. In those cases, a modification of the filter channel or adjacent ductwork may be necessary.

Filters should be changed regularly and not allowed to become heavily loaded. One recent study suggested that filters can become sinks for indoor air contaminants and can actually contaminate the airstream they are supposed to clean. Therefore, filters may need to be replaced more frequently than generally accepted.

13.3.4 Humidification

Relative humidity (RH) in an occupied space should be maintained between 30% and 60% to help minimize the growth of pathogenic and allergenic microorganisms. RH can exceed 60% during the summer especially in the south. High RH can promote the growth of fungal species in occupied spaces and low velocity ducts and plenums. Conversely, it is not unusual for RH in the winter to be below 20%. RH below 20% can cause drying and irritation of the mucous membranes and have been associated with an increasing potential to spread respiratory illnesses. This is due to aerosols produced by human sneezing and coughing, and by water sprays are desiccated to the respirable particle size range in low humidity environments[2].

Proper installation and O&M of a humidification system are very important. Care should be exercised in the selection of humidification systems. Cool mist, water spray and ultrasonic systems have a greater potential than steam systems to emit microbials and particulates into the airstream if not properly designed and maintained. Steam type humidification systems should not use boiler steam because boiler treatment chemicals can be aerosolized. All types of humidifiers need aggressive and proactive maintenance.

13.3.5 Condensate Pans and Standing Water

Condensate pans should be self-draining. Pans and fan chambers should not have any standing water. Condensate pans can be at risk to become contaminated with microbial growth due to inadequate filtration and standing water. Pathogenic and allergenic microorganisms can flourish in the pans. Proper design and maintenance of the pans can help minimize the potential for microbial contamination. It is likely that the building engineer will have to modify existing condensate drainage systems to avoid any standing water or drying out of traps.

13.3.6 Ductwork

Ductwork should be periodically examined and cleaned, if necessary, to maintain system hygiene. Interior bare duct and insulation can accumulate dirt and moisture, causing conditions favorable for microbial contamination. Bare galvanized ductwork has some degree of biocidal activity due to the zinc contained in the galvanized metal. Internal insulation, on the other hand, most likely will offer sites for microbial contamination if dirty and wet.

Duct cleaning should be performed by reputable firms which use and are well experienced in the procedures recommended by National Air Duct Cleaners Association (NADCA) Standard 1992-1. Since air ducts can be contaminated with microbial and particulate contaminants, the duct-cleaning project should be well planned and use techniques that minimize the chances of contamination of the occupied space with debris from the duct.

13.3.7 Energy Conservation Systems

The building engineer should evaluate whether any installed economizer cycle is working properly. Economizer cycles allow more ventilation without incurring an energy penalty. Additionally, it is important to check the energy management computer or other HVAC control systems for proper operation. Optimum start/stop times should be carefully selected to provide preoccupancy ventilation according to the recommendations of ASHRAE Standard 62-1989.

13.3.8 Precautions to be Taken During Maintenance And Renovations

During the maintenance and renovation of HVAC systems, special precautions must be taken to maintain adequate IAQ and thermal conditions not only in the maintained or renovated space but also in adjacent spaces. Segregation and negative pressurization of the construction area(s), proper scheduling of work, dust mitigation techniques, use of low volatile organic compound (VOC)-emitting materials, and use of temporary filters in the adjacent return systems can avoid many IAQ problems that occur during renovations. Proper planning and management techniques are crucial for maintaining IAQ in a building.

13.4 SAMPLING OF IAQ PARAMETERS

While the merits and degree of IAQ sampling may be debated among IAQ practitioners, many building managers will continue to commission IAQ sampling either in response to complaints, or as part of a proactive IAQ program. Therefore, it is important for building managers to be familiar with some of the more commonly measured indoor air contaminants.

Proactive sampling programs offered by some IAQ consultants typically sample contaminants in occupied areas at least semiannually, usually in both the winter and summer. Areas sampled are randomly selected but are representative of the majority of the space on a particular floor. Each floor of a building is sampled if there is a significant difference in occupancy and type from floor to floor. At a minimum, one area served by each individual AHU is sampled.

When commissioning IAQ sampling, the building manager should ensure that the sampling is performed by a qualified individual (either on staff or an outside consultant) who is experienced in measuring IAQ parameters. A Chain of Custody should be required for all samples submitted to laboratories for analysis. The sampling report should contain at least the following data:

- Description of the area sampled
- Its associated occupancy at the time
- The time of day
- Outside reference levels (if applicable)
- Sampling methodology

- Applicable standard or guideline value
- Actual measured data
- Data from the previous semiannual inspection

A sampling example for CO_2 is illustrated below:

Area	Occupancy	Time	Outside value (ppm)	Sampling method	Guideline value [b] (ppm)	Current measurement (ppm)	Previous inspection (ppm)
Internal audits	10	11 am	350	(a)	800	700	650
Payroll	25	11 am	350	(a)	800	500	600
	2	11 am	350	(a)	1000	375	375

lobby
(a) direct reading infrared CO_2
(b) should be explained in report

Listed below are example parameters that may be measured by IAQ sampling firms and consultants. This listing is made solely to make the building manager aware of some of the parameters measured by IAQ consultants. Development of specific sampling programs and protocols can only be made after a comprehensive IAQ survey and assessment is performed by an IAQ specialist.

13.4.1 Carbon Dioxide

Carbon dioxide (CO_2) is one the most important IAQ parameters measured in buildings. CO_2 is nontoxic in levels usually found in office buildings and, therefore, not a concern in itself. However, it is used by many IAQ practitioners as an indicator of ventilation in a space. The outside concentration of CO_2 typically ranges between 350 ppm and 500 ppm. Indoor concentration is usually a function of the number of persons in the space and ventilation (quantity and effectiveness). ASHRAE Standard 62-1989 recommend that indoor CO_2 concentrations be below 1000 ppm. The standard also assumes 7 persons per 1000 ft^2 (actual area, not rentable). OSHA recommends that CO_2 concentrations not exceed 800 ppm in offices[3]. Other agencies such as the U.S. Air Force sets a lower standard of 600 ppm[4]. Regardless of what indoor concentration standard is used it is important to also measure the outside concentration at the same time.

When and where to measure CO_2 concentrations must also be carefully selected as well as the method used to measure the CO_2. OSHA recommends that the sampling be performed late in the morning (11:00 am) and late in the afternoon (3:30 am) at a height of 3 feet to 5 feet above the floor (at the breathing zone for most seated individuals). It recommends that the sampling be performed in occupied areas but in locations where air mixing is least effective such as corners in offices.

There are two common methods to measure CO_2 : direct reading colorimetric detector tubes and real time, direct reading infrared detectors. When using colorimetric tubes, at least two samples should be taken in each location and their values be averaged. OSHA has recommended the following schedule of measurements to ensure a 95% confidence that the concentration of CO_2 measured is below 800 ppm.

Number of Samples	Average CO_2 (ppm)
2	670
3	695
4	710
5	720
6	725
7	730

When using direct reading CO_2 infrared detectors, the +/- accuracy of the meter should be known and used. For example, if the meter has an accuracy of +/- 25 ppm at full scale and the CO_2 reading is 775 ppm, the actual CO_2 concentration could be between 750 ppm and 800 ppm. Regular calibration of instruments is essential.

13.4.2 Temperature and Relative Humidity

Temperature can have a significant effect on the perception of IAQ. Occupants are more likely to associate warmer air with "stuffy air." ASHRAE Standard 55-1992: *Thermal Environmental Conditions for Human Occupancy*, specifies the combinations of indoor space environment and personal factors that will produce thermal environmental conditions acceptable to 80% or more of the occupants within a space. It should be recognized that because of the differences in the way people perceive temperature it is impossible to satisfy 100% of the building occupants at all times.

RH is also very important. In winter, relative humidity in offices and homes can become very low. It is not unusual to find RH below 20% in the northern latitudes. Occupants will feel colder when the RH is low. Mucous membranes tend to dry out and thereby contribute to an increase in susceptibility to respiratory illnesses. Additionally, as people sneeze and cough, the expelled particles desiccate very rapidly in low RH and, therefore, become lighter and tend to remain airborne longer. Combined with low ventilation rates, this can increase the likelihood of spreading cold and flu[2]. In summer, RH can be higher than 60% that can cause fungi to form colonies on building surfaces. Some of these fungi can produce a number of reactions, including allergies, hypersensitivity pneumonitis and fungal infections in sensitive people.

ASHRAE Standard 55-1992 provides a range of acceptable operating temperatures and concurrent RH that will satisfy 90% of the building population. Using the recommendations of the standard, space comfort criteria should be developed for each building. For illustrative purposes only, an example of comfort criteria is shown below.

Season Temperature (°F)		Relative Humidity (%)
Winter	(70-74)	35-50
Summer	(73-76)	45-55

13.4.3 Formaldehyde

Formaldehyde is a VOC used in the manufacturing process of thousands of products. It can offgas from building materials such as particle board and urea formaldehyde foam insulation. Formaldehyde can also be produced by the incomplete combustion of hydrocarbon fuels. High indoor concentrations of formaldehyde can cause people to experience burning sensations in their mucous membranes (eyes, nose, etc.). Chronic exposure to high concentrations of formaldehyde can destroy mucous linings and increase the susceptibility to respiratory diseases. The formaldehyde offgas rate of building materials varies proportionally with temperature and RH. Indoor concentrations are also affected by the dilution of the ventilation system. The odor threshold for formaldehyde is around 0.075 ppm. But even this concentration may not be low enough for hypersensitive individuals.

13.4.4 Carbon Monoxide

Carbon monoxide (CO) is a colorless and odorless gas that can be lethal. It is a byproduct of incomplete combustion (insufficient oxygen when burning fossil fuels). High indoor concentrations (greater than 35 ppm) of CO can cause headaches and nausea. OSHA limits the CO concentration in a workplace to 50 ppm for a one-hour exposure. In the absence of indoor combustion appliances, the indoor CO concentration should not exceed the outside concentration. Some of the potential sources of CO include enclosed garages, buildings built on top of garages (CO can seep through cracks and other pathways into the building), and buildings whose outside air intakes are adjacent to garages.

13.4.5 Ozone

Ozone (O_3) is a form of oxygen that is highly reactive. It is commonly used as a biocide in water systems. O_3 is commonly produced outdoors by electrostatic reactions (thunderstorms) and photochemical reactions (smog). Significant indoor sources are electrostatic air cleaners, malfunctioning motors, certain photocopiers and laser printers. Exposure to significantly high concentrations of ozone (greater than 0.1 ppm) can cause mucous membrane and respiratory irritation, and headaches.

13.4.6 Nitrogen Dioxide

Nitrogen dioxide (NO_2) is a byproduct of combustion that can affect the respiratory system. It is a pungent gas commonly produced indoors by gas ranges, kerosene heaters, wood stoves and cigarette smoking. Typical concentrations in offices will usually be less than or equal to the outside concentration unless there are combustion sources in offices.

The National Primary Ambient Air Quality Standard (NPAAQ) for outside air specifies that the NO_2 yearly average be no higher than 0.055 ppm. The American Council of Governmental Industrial Hygienists (ACGIH) threshold limit value (TLV) (based on time-weighted average [TWA] for 8 hours) for industrial workplaces is 3 ppm. However, sensitive populations such as people with asthma and young children can be affected by NO_2 concentrations as low as 0.15 ppm. Indoor NO_2 concentration should be less than or equal to the outside concentration when no combustion sources are present in the space.

13.4.7 Sulfur Dioxide

Sulfur dioxide (SO_2) is another combustion byproduct of sulfur-containing fuels that can cause respiratory problems. Indoor sources are usually kerosene heaters. SO_2 is highly soluble in water and can be readily absorbed by the mucous membranes forming sulfuric acid. People with asthma are especially sensitive to SO_2 in concentrations at or above 1 ppm. Sulfur dioxide is highly reactive and, therefore indoor concentrations will usually be lower than outdoor concentrations.

13.4.8 Biocontaminants

Bacteria, viruses and fungi are common in the indoor air. Most indoor air microbes (bacteria and fungi) are harmless to healthy adults. However, there are some species, when amplified in the indoor environment, can cause allergies and other diseases to susceptible populations. Occupants of a building are a common source of bacteria that are shed from skin. More worrisome are airborne bacteria associated with respiratory diseases. Such bacteria include *Legionella pneumophila* (legionnaire's disease), *Pseudomonas* species (humidifier fever) and *Mycobacterium* (tuberculosis). Certain fungal species can cause allergic reactions in some people such as *Aspergillus flavus*, *Stachybotrys atra*, *Aspergillus fumagatus* and certain *Penicillium sp.* These species and others can amplify in the HVAC system and other building materials and can be spread throughout the building by the HVAC system.

Common methods to measure the biocontaminant load in a building involve bulk or wipe samples plates, and impact plate samplers. The bulk or wipe samples are used in a qualitative sense to determine if any pathogenic species are present in the HVAC system and on other building surfaces. Plates containing a nutrient are brought momentarily in contact with the surface to be analyzed. The plates are then sent to a laboratory for incubation and analysis. The second method involves measuring the number and type of viable particles in the air. Both bacteria and fungi can be measured in colony forming units per cubic meter (cfu/m^3). A known quantity of air is drawn through a sieve impactor into contact with an agar plate. The plates are then sent to the laboratory for incubation and analysis.

There are currently no legal standards for biocontaminants in the indoor air. However, certain organizations and individual researchers have suggested guidelines that are generally accepted as reasonable. Common factors to most of the proposed guidelines include that: (a) outside and indoor fungi types should be similar, (b) indoor fungi level should be lower than the outside level (indoor to outdoor cfu/m^3 ratio of less than 0.5), and (c) gram-positive bacteria should be the predominant indoor bacteria.

13.4.9 Respirable Particles

Airborne respirable particles are generally accepted as those particles below 5 microns in diameter. Respirable particles can be breathed in deeply into the lungs past the body's normal defenses. Some of the particles can be viable such as fungal spores, bacteria and viruses. Other particles can be nonviable such as asbestos, fiberglass, and airborne particulate matter from ETS. Currently, the EPA has set a standard only for outside concentration of particles (PM_{10}) as 10 microns. High concentrations may indicate the need for upgraded filtration and/or better housekeeping.

13.5 OCCUPANT IAQ COMMUNICATION PROGRAM

Effective communication with occupants is a crucial element of an IAQ program. It is important to have several ways of communication to relay the message that the building has a proactive IAQ program. The following three tiered communication programs are examples of an approach that may be used by a building manager having a proactive IAQ program. Each tier has its advantages and, when used together, compliment each other.

The first is publishing details of the proactive IAQ program in a building newsletter. The article should be concise and inform occupants that building management has a proactive IAQ program to enhance the quality of the air they breathe. It should list key contacts with their phone numbers for those individuals who would like more information or have concerns. Of course, tenant representatives should be solicited for their input prior to building-wide distribution of the article.

The second involves a lunch or breakfast seminar on the building IAQ program. This seminar consists of a professional quality presentation of the proactive program by building management (and IAQ consultant, if used). The presentation should offer specifics of the proactive steps taken to improve the indoor environment, as well as offer advice to occupants on how they can help building management in ensuring adequate IAQ. The presentation should conclude with a general question/answer period.

The final is individual employee and tenant presentations. This is most effective when addressing immediate IAQ concerns. The substance of the presentation will vary according to circumstances. In response to a complaint, the presentation should address the immediate actions taken to resolve the problem/concern and what IAQ enhancement measures are being implemented. When the presentation is for general information, the outline should be similar to the IAQ seminar described above.

An important element of any proactive IAQ program is the maintenance of building systems that can affect IAQ. Such building systems include, but are not limited to, HVAC components, cooling towers, potable water systems and air conveyance systems. The goal of the maintenance program should be to ensure service life, capacity, availability, and hygiene of the systems. Maintaining systems for IAQ may, in some cases, require adherence to a more stringent standard of care than is presently performed.

Table 13-1 shows a sample maintenance schedule for some of the building systems that can affect IAQ. When a manufacturer recommendation conflicts with an IAQ maintenance schedule, the stricter of the two should apply. This example schedule is intended solely to make building operating personnel aware of some of the elements that can make up an actual IAQ maintenance schedule. It does not list all systems that can affect IAQ nor list all required maintenance activities and schedules that may be necessary to maintain adequate IAQ in a specific facility. Therefore, it should not be applied to specific buildings and its systems. Specific maintenance schedules can only by developed after a full and thorough assessment is made of all building systems and a comprehensive IAQ building assessment is performed.

Table 13-1. A Sample Maintenance Schedule.

ITEM	MAINTENANCE ACTIVITY	SCHEDULE
Prefilters and main filters	Inspect for breakthrough, contamination and integrity of fit	Weekly
Prefilters	Replace	Earliest of: • Every 4 months • When indicated by pressure-drop

Table 13-1. A Sample Maintenance Schedule (*Cont'd*).

ITEM	MAINTENANCE ACTIVITY	SCHEDULE
Main filters	Replace	Earliest of: • Every 12-18 months • When indicated by pressure-drop
AHU* internal chamber	Visual inspection	• Daily (large units located in mechanical equipment rooms) • Monthly for others
Condensate pans	Clean and sanitize	Monthly
Condensate traps	Check for seal	Weekly in winter
Humidifiers	Check for proper operation	Weekly
	Check blowdown operation	As recommended by manufacturer
Ductwork	Inspect for contamination, integrity of insulation	Representative section every 6 months (especially first 10 feet after AHU)
	Clean	When indicated by visual examination
HVAC controls	Adjust and calibrate	Semiannually (start of heating and cooling season)
Outside air intakes, dampers and actuators	Inspect and test for proper operation	Every 3 months
Cooling and heating coils	Power wash	Power wash yearly
Cooling towers, ornamental fountains and potable water systems that can be aerosolized	Check for proper water treatment	As indicated by water treatment manufacturer

Table 13-1. A Sample Maintenance Schedule (*Cont'd*).

ITEM	MAINTENANCE ACTIVITY	SCHEDULE
Air plenums	Inspect for cleanliness, integrity, blockage and water damage	Semiannually
Ventilation system	Check capacity and air balance	Every 5 years or after renovation

*AHU: Air-handling unit.

13.6 HANDLING IAQ COMPLAINTS

Handling IAQ complaints from occupants in a building is perhaps the most critical element in helping to diffuse a potentially serious condition. All complaints or inquires expressing concern about the indoor environment should be taken seriously and handled as a top priority. While dispatching a mechanic or an HVAC technician for routine thermal complaints is appropriate, specially trained individuals should respond to IAQ concerns. These individuals should have basic knowledge in IAQ fundamentals. Preferably, they should have attended one of the many available IAQ training courses given by various organizations. The Indoor Air Division of the EPA can provide a list of such courses.

Proper documentation of IAQ complaints is very important. Extreme care should be taken in all aspects of the investigation. Complaint investigation forms, similar to the forms in the EPA's Building Air Quality Guide may be used. It is just as important not to overreact to an IAQ complaint as underreact. Showing up with a number of technicians, consultants, etc. can give the impression that there is a dangerous condition present. Usually, prompt and courteous attention by a knowledgeable individual with good communication skills will help restore confidence in management.

The worst thing building management can do is to ignore IAQ complaints or not take them seriously. Besides souring tenant/landlord relations, the costs to resolve an IAQ complaint that is ignored will rise exponentially with time as more parties get involved. Once the press, lawyers and government investigators get involved, the costs to building owners and operators will be many times what it could have been if resolved at the time of the first complaint.

13.6.1 Handling First Contact

Usually, an IAQ complaint will be called in either by a tenant representative or the individual complainant. The person receiving the call should obtain from the individual his or her telephone number, the exact nature of the complaint, the location of the problem area, and when the symptoms first appeared. The complainant should then be informed that an IAQ investigator will contact him/her by the end of the next business day to make an appointment to visit and obtain further information. The person receiving the complaint should record the information on a complaint log and immediately contact an IAQ investigator. He or she should also note the date and time call came in and when an IAQ investigator was notified.

13.6.2 Interviewing Complainant

When interviewing occupants, it is important to use a questionnaire. This form will assist the IAQ investigator in gathering all pertinent information in a way that will not bias the answers and will ensure that if an IAQ consultant needs to offer assistance later on, important information was not overlooked. The occupant questionnaire form of the EPA's Building Air Quality Guide may be used. The occupant interview should be performed in the occupant's office or another private location. The interviewer should explain that answers to the questions asked will be of assistance in finding out the cause of the problem.

13.6.3 Gathering Physical Data

The next step is to gather important physical data that, in conjunction with the answers to the questionnaire, will help forming a hypothesis for testing. What to measure, if any, depends on the nature of the complaint and answers to the questionnaire. At a minimum, temperature, relative humidity and CO_2 should be measured and recorded. Also, proper functioning of the HVAC system and contaminant sources and pathways should be checked in every complaint area. Forms in EPA's Building Air Quality Guide may be used to document the gathering of physical data. When performing CO_2 measurements, the procedure outlined in the OSHA IAQ Rule should be used.

13.6.4 Formulating A Hypothesis

Once all the physical data and interview information are gathered and analyzed, a hypothesis of the cause of the complaint can be developed. The cause of the complaint may be obvious in some cases and not at all apparent in others. One should look for sources of problems such as the HVAC system, office equipment, cleaning chemicals, other contaminants, and pathways of contaminants into the problem areas. On the other hand, one should not overlook other causes of environmental stress such as noise, temperature and lighting.

13.6.5 Develop and Implement an Action Plan

Once a hypothesis is developed, it should be tested by developing and implementing an action plan. In most cases involving malfunctioning of HVAC equipment, correction of the problem will quickly test the hypothesis. In systemic or multicause cases, the hypothesis testing may take days, weeks or even months (such as when IAQ symptoms are hypothesized to be caused by overcrowding, an underdesigned ventilation system or installation of new furnishings offgassing an irritating contaminant). Meanwhile, the occupants should be well informed of the efforts being made during the hypothesis testing period.

13.6.6 Soliciting Feedback on Resolution

Once the action plan has been implemented, the IAQ investigator should solicit feedback from the complainant. If the complainant indicates that the problem has been resolved, clear documentation of this fact should be made. If the problem persists, the hypothesis may have been incorrect or other causes have occurred subsequent to the initial investigation. If more information is available, a new hypothesis should be developed. Even if the immediate problem has been resolved, one should check back with the complainant periodically to see how things are progressing.

13.6.7 Calling in an IAQ Consultant

When in-house attempts to resolve an IAQ problem fail, the IAQ investigator may need to solicit help from an IAQ consultant. The IAQ investigator should gather all documentation available for review by the consultant, and should use the recommendations in the EPA's Building Air Quality Guide in soliciting help from IAQ consultants.

13.7 REFERENCES

[1]U.S. EPA. "Building Air Quality, A Guide for Building Owners and Facility Managers."

[2]Burge, H.A., et al. "Indoor Air Pollution And Infectious Diseases," p. 277, *Indoor Air Pollution, A Health Perspective*, edited by J. Samet, M.D., Johns Hopkins University Press.

[3]Federal Register. "Indoor Air Quality, Proposed Rule," April 5, 1994, pp. 15968-16039.

[4]USAF. "Guide For Indoor Air Quality Surveys"

[5]Etkin, D. "Biocontaminants in Indoor Environments," Cutter Information Corporation.

[6]Godish, T. "Indoor Air Pollution Control," Lewis Publishers.

14

Biocontaminant Control For Acceptable IAQ

Philip R. Morey, Ph.D., CIH
Vice President, Director of Microbiology
AQS Services, Inc.
Atlanta, Georgia

14.1 INTRODUCTION

Microorganisms should not be allowed to grow in buildings and in heating, ventilating and air-conditioning (HVAC) systems because of their potential adverse health effects. In residences, the presence of moisture and resulting mold growth has been associated with both respiratory and non-respiratory diseases[1-3]. The growth of microorganisms in HVAC systems[4-7] or in the water systems of buildings[8,9] has been associated with the occurrence of diseases including hypersensitivity pneumonitis, humidifier fever, stachybotrytoxicosis, Pontiac Fever and Legionnaires' Disease.

The growth of molds, bacteria and mites in buildings is primarily controlled by moisture availability, and to a lesser extent, by the presence of nutrients such as dirt and debris. Proper design, and operation and maintenance of HVAC systems can reduce the growth of microorganisms in buildings.

14.2 MOISTURE AND GROWTH OF MICROORGANISMS

Microorganisms grow on or in construction and finishing materials in both occupied spaces and in HVAC system components. The amount of moisture available in the substrate primarily determines what kinds of microorganisms can grow. An equilibrium relative humidity (ERH) of 65% defines the driest conditions compatible with the growth of molds on almost all materials[10]. ERH is defined as the vapor pressure exerted by water in a substrate expressed as a percentage of the saturation vapor pressure of pure water at the same temperature and pressure. An ERH of 65% that potentially allows some molds to grow is achieved if the relative humidity in room air is consistently maintained at 65% so that the moisture content of materials (potential substrates) in the room comes into equilibrium with the moisture in the air.

There are at least 100,000 different kinds of molds. Xerophilic molds such as *Wallemia*, *Eurotium* and *Aspergillus versicolor* are capable of growing under the most limiting conditions of water availability (65% < ERH < 85%). Hydrophilic molds such as *Stachybotrys* and *Fusarium* grow on substrates where the ERH consistently exceeds 90% such as wet gypsum board or wet surfaces such as in air-handling unit (AHU) drain pans. Microorganisms such as bacteria and yeasts grow in liquid water. Gram-negative bacteria grow on wet surfaces of cooling coils[11]. Humidifiers or water spray systems that emit water droplets from a sump with recirculated water are especially prone to contamination by Gram-negative bacteria such as *Pseudomonas* and *Flavobacterium*.

The water in well maintained cooling tower reservoirs normally harbors bacteria at concentrations that may approach 10,000 or 100,000 per milliliter. *Legionella*, a Gram-negative bacterium that can grow in cooling tower water, can cause building-related Legionnaires' Disease providing that a water droplet (or droplet nucleus) containing the living bacterium enters a building opening (e.g., outdoor air inlet) and eventually is deposited in the lungs of the susceptible person. *Legionella* may also grow in potable water systems especially in stagnant, tepid conditions and this bacterium can be transferred to the lungs by water droplets from outlets such as shower heads.

The aeroallergens of the house dust mite are thought to be one of the most important causes of asthma especially in residential environments[12]. Mites grow optimally in extended surfaces, fleecy materials such as carpets and bedding where the concentration of nutrients (human skin scales) is unlimited and the relative humidity is approximately 75% to 80%. Mites, however, can survive in fleecy materials where the relative humidity is above 55%[12].

14.3 MOISTURE AND BIOCONTAMINANT CONTROL

Molds, bacteria and mites grow in buildings and in HVAC systems primarily because of availability of moisture. Moisture control in buildings involves proactive design, operation and maintenance considerations. If moisture is not controlled, extensive (more than approximately 30 square feet[13]) visible mold contamination can result in a building unfit for human occupancy[14]. Principles of moisture control necessary to prevent the growth of molds and other biocontaminants are described in the following sections.

14.3.1 Room Relative Humidity

A relative humidity in a room consistently maintained at 65% (ERH = 65%) is equivalent to the lowest substrate moisture content capable of supporting mold growth in construction and finishing materials. As the relative humidity in the room air increases above 65%, the likelihood of mold growth increases[15]. Preventing the relative humidity from consistently exceeding 60% in indoor air in most cases provides an adequate safety factor for reducing the moisture content in construction and finishing materials to levels that will not support mold growth. On the other hand, a relative humidity of 60% does not guarantee the absence of mold growth when a room surface is cold. The surface relative humidity on a cold ceiling or wall may be 70% or 80% or even higher when the relative humidity in the middle of the room is 60% or less[16].

14.3.2 Moisture in Construction and Finishing Materials

The constant wetting of materials in buildings results in growth of microorganisms. The soiling of construction and finishing materials makes them more hydrophilic (take up moisture more rapidly), and this allows molds to grow under more moisture limiting conditions in comparison to equivalent clean materials. Proactive actions for preventing biocontamination in construction and finishing materials include the following:

(a)   Flood-damaged materials should be dried as rapidly as possible (preferably in 24 hours[17]) to an ERH of less than 65% that will not support mold growth.

(b)   The moisture present in some materials (e.g., wet plaster, wet wood, etc.) used in new construction must be reduced by drying to an equilibrium moisture content that does not exceed 65% on its sorption isotherm. An equilibrium moisture content is defined as the amount of moisture in a material maintained in equilibrium with an atmosphere at a certain relative humidity.

(c)   The soiling of construction and finishing materials should be prevented.

14.3.3 Water Droplets from Outdoors

Droplets of water from cooling towers, fog, rain and snow should not enter building openings especially HVAC system outside air inlets. Water droplets from cooling towers potentially contain *Legionella*. The moisture associated with impaction of water droplets into HVAC systems provides a niche for mold growth. Proactive measures for preventing biocontamination include the following:

(a)   Position HVAC system outside air inlets and other building openings at least 50 feet from cooling towers and evaporative condensers. Prevailing winds and air flow patterns around the buildings should not result in entrainment of water droplets in building openings. The practice of placing cooling towers together with HVAC system outside air inlets and other building openings within the same architectural fence should be avoided.

(b)   HVAC system outside air inlets should be located sufficiently above the ground and above roofs to prevent entrainment of snow and rain. The consistent entrainment of fog in HVAC system outside air inlets will result in mold growth in filters.

14.3.4 Moisture Generated in HVAC Systems

Moisture conditions in HVAC systems that result in bacterial and mold growth are unacceptable. The aerosolization of microbicidal chemicals into operating HVAC systems for purposes of controlling microbiological contamination is also unacceptable. Proactive measures for preventing biocontamination in HVAC systems include the following:

(a)   Dehumidification (cooling) coils should be designed in such a way that water droplets will not be carried over onto downstream surfaces. Humidifiers emitting water vapor should be designed so that sufficient distance is provided to allow moisture to fully vaporize to prevent wetting of all downstream surfaces. Water droplets emitted by water spray systems or humidifiers should be completely contained by demister plates and should not contain microbial contaminants above those normally present in potable water.

(b)   Drain pans under cooling coils, air washers and humidifiers should be designed to avoid water stagnation. The presence of biofilm, slime and mold in HVAC equipment should be prevented by frequent thorough cleaning[18]. Physical removal of biofilm, slime and mold is required (disinfection alone is insufficient) because dead microbial components can still cause allergic or toxic diseases.

14.3.5 Porous Insulation In HVAC Systems

Porous insulation used to line the airstream surfaces of HVAC equipment provides a locus for the accumulation of dirt and debris. Dirt and debris are hydrophilic and the insulation on the airstream surfaces in mechanical cooling systems (relative humidity of air consistently exceeds 65%) thus provides a niche for mold growth[19,20]. The mold growing on porous insulation or on the resin in porous insulation[20] unlike moldy debris on a hard surface such as a sheet metal cannot be removed by duct cleaning.

Proactive measures for preventing biocontamination on HVAC system insulation include the following:

(a)   Porous insulation should not be used to line the airstream surfaces of HVAC components where wetting is likely such as in the immediate vicinity of cooling coils, water spray systems, humidifiers, or other sources of water. Thus, plenums housing cooling coils, water spray systems, drain pans and humidifiers should not be lined on the airstream surface with porous insulation. Porous insulation may, however, be used in these HVAC components provided that the insulation is separated from the moisture sources by a barrier that is both airtight and watertight.

(b)   Because of possible mold growth[19,20], the use of porous insulation should be minimized on the airstream surfaces in mechanical ventilation systems where the relative humidity of conditioned air consistently exceeds 65%.

14.3.6 Condensation in Building Assemblies

Condensation can occur in the envelopes of buildings when moist air encounters a cold surface. In air-conditioned buildings in warm humid climates, condensation occurs when moist air infiltrates into the envelope and encounters a cold surface on the side of the wall facing the occupied space[21]. In cold climates, condensation can occur when moist indoor air flows out through porous construction and encounters cold surfaces toward the outside of the building[22]. Mold growth can occur when condensation moisture remains trapped in the envelope.

Vapor diffusion and air retarders can be used to prevent moisture accumulation in building envelopes. In predominantly cold climates, the retarders are located toward the interior of the envelope. In warm humid climates, the retarders are located toward the exterior of the envelope.

14.3.7 Potable Hot Water System

Proactive guidance on control of *Legionella* in potable hot water systems is well known[8,9,23]. Hot water tanks should be sized to satisfy the building's needs. Conditions of water stagnation in tanks and pipework should be avoided. For example, oversizing of tanks can result in tepid water which may provide the ideal temperature range (85°F to 105°F) for *Legionella* growth.

Hot water service should be designed to allow for periodic pasteurization. Thus, hot water tanks should be able to be periodically heated so that their contents reach 140°F, and deliver water at a minimum temperature of 131°F to all outlets. Safeguards, however, must be employed to prevent scalding during periodic pasteurization (disinfection) of the system to kill *Legionella*. Hot water service should be physically isolated from (or insulated from) cold water systems to reduce heat transfer. This is intended to reduce the likelihood of *Legionella* colonization in cold water systems.

14.4   EXAMPLES OF FAILURE
IN PREVENTING BIOCONTAMINATION

A considerable amount of information is available on prevention of biocontamination in buildings[24]. Failure to prevent biocontamination can, in worst cases, result in significant occupant exposure to bioaerosols and possible adverse health effects. In the three examples that follow, amplification of microorganisms occurred in buildings or building systems. Disease was reported in two of the buildings. In the third building, the owner recognized the potential for bioaerosol exposure and took necessary action to prevent occupant exposure from biocontaminants that were present.

14.4.1 Legionnaires' Disease

Two cases of pneumonia caused by *Legionella pneumophila* serogroup No. 1 were reported at a facility where employees worked both outdoors and indoors (offices). Because the facility did not have a cooling tower, an environmental search for *Legionella* was concentrated in the facility's potable water system. *Legionella* was not detected in mains water entering the facility. However, substantial concentrations of *Legionella pneumophila* serogroup No.1 were detected in water from an outside storage tank (located without protection in direct sunlight), in stagnant water in water spray hoses used for dust suppression, and in the office hot-water tank (refer to Table 14-1). The water temperature in much of the cold water system often exceeded 75°F and water temperatures in the hot water service were tepid (90°F to 110°F).

Disinfection of the hot water service was achieved by pasteurization (refer to section 14.3.7 of this chapter). The cold water system was disinfected by chlorination (10 parts per million [ppm] to 20 ppm free residual chlorine)[8]. *Legionella* growth in this facility could have been prevented by avoiding both stagnation in storage systems and tepid conditions in both hot and cold water services.

14.4.2 *Aspergillosis*

Several cases of aspergillosis occurred in a bone marrow transplant unit in a medical center. This disease is caused by infection (growth) of *Aspergillus fumigatus* or *Aspergillus flavus* in the lungs. Outbreaks of aspergillosis in medical centers are usually associated with entrainment of spores in organ transplant patient areas from one or more diverse sources including dusts in the outside air, HVAC system dusts, and dust from interior renovations.

Inspection of the AHU serving the bone marrow transplant unit

Table 14-1. Growth of *Legionella* in a Potable Water System.*

Description and Location of Water Sample	Temperature (°F)	*Legionella* per Milliliter
Mains water	65	None detectable
First draw from bottom tap of storage tank	75	100 *Legionella pneumophila,* serogroup 1
Stagnant water in hose line	110	1600 *Legionella pneumophila,* serogroup 1
First draw from hot water tank	100	600 *Legionella pneumophila,* serogroup 1

* *Legionella* determined by culture on buffered charcoal yeast agar containing antibiotics selective for this organism (reference 25 in text). The limit of detection was approximately one organism per milliliter of sample.

showed that both low efficiency and final filters were physically wet. Low efficiency filters were wet because of impaction of fog on filter media. Final filters were wet because of water droplet carryover from cooling coils.

Culture analysis of filters showed that the filter media itself was a growth site for thermotolerant *Aspergillus fumigatus* that had caused patient infection (refer to Table 14-2). This mold grows well in substrates that are almost water saturated (ERH > 90%). This epidemic among high susceptible bone marrow transplant patients could have been avoided by preventing fog entrainment in the AHU (refer to section 14.3.3 of this chapter) and by the absence of water droplet carryover from cooling coils at design conditions (refer to section 14.3.4 of this chapter).

14.4.3 Mold on Chronically Wetted Gypsum Board

Because of chronic roof leaks, gypsum board walls in one area of an office building periodically became saturated with water. Visible mold was observed on at least 300 ft² of gypsum board within several weeks.

Table 14-2. *Aspergillus* **Growth in HVAC Components.***

Sample Description	*Aspergillus* per Gram of Sample
Low efficiency filters	Greater than 10,000 *Aspergillus fumigatus*
Final, high efficiency (95%) filters	100 to 1000 mostly *Aspergillus fumigatus;* some *Aspergillus flavus*
Debris in plenum housing final filters	1000 *Aspergillus fumigatus*

* Samples processed by dilution culture on malt extract agar and incubated at 37°C (98.6°F).

Laboratory analysis of pieces of moldy gypsum board revealed the presence of many culturable fungi including several mycotoxin producers such as *Stachybotrys chartarum, Penicillium brevicompactum, Penicillium aurantiogriseum,* and *Aspergillus versicolor* (refer to Table 14-3).

Moldy gypsum board was removed from the building under containment similar, but not identical, to procedures used during asbestos abatement[13,26,27]. Air sampling performed during removal of moldy gypsum board showed that concentrations of airborne *Penicillium-Aspergillus* and *Stachybotrys* spores within the containment work area were approximately 10 million/m³ and 10 thousand/m³, respectively (refer to Table 14-3).

By requiring that microbial remediation be performed under containment, the owner prevented dissemination of spores into clean areas of the building. When exposure to bioaerosols of the type in Table 14-3 was anticipated, all persons involved in remediation of moldy gypsum board were provided with respiratory and other personal protective equipment[26]. While not all floods and water damage in buildings can be prevented, it is nevertheless important in preventing mold growth to dry water-damaged materials as quickly as possible, preferably within 24 hours (refer to section 14.3.2 of this chapter).

Table 14-3. Mold on Chronically Wetted Gypsum Board.

Sample Description	Concentration and Types of Mold
Moldy gypsum board, side facing occupied space	600,000 culturable* molds/in^2 *Stachybotrys chartarum* (34%) *Penicillium chrysogenum* (31%) *Verticillium lecanii* (16%) *Penicillium brevicompactum* (8%) *Penicillium aurantiogriseum* (7%) *Aspergillus versicolor* (4%)
Inside (within wall) face of another moldy gypsum board	1,300,000 culturable molds/in^2 *Penicillium chyrosgeum* (88%) *Stachybotrys chartarum* (8%) *Aspergillus versicolor* (3%) *Cladosporium cladosporioides* (1%)
Airborne spores** during removal of moldy gypsum board	10 million *Penicillium-Aspergillus* spores/m^3 10,000 *Stachybotrys* spores/m^3

* Molds cultured on malt extract agar.
** Spores collected using Burkard spore trap operating at a flow rate of 0.01 m^3/ minute.

14.5 REFERENCES

[1]Brunekreef, B., D. Dockery, F. Speizer, J. Ware, J. Spengler, and B. Ferris, "Home Dampness and Respiratory Morbidity in Children," *Am. Rev. Resp. Dis.*, 140, pp. 1361-1367, 1989.

[2]Dales, R., R. Burnett, and H. Zwanenburg, "Adverse Health Effects in Adults Exposed to Home Dampness and Molds," *Am. Rev. Resp. Dis.*, 143, pp. 505-509, 1991.

[3]Miller, J., "Fungi and the Building Engineer," IAQ '92, Environments for

People, pp. 147-159, American Society of Heating, Refrigerating and Air-Conditioning Engineers, Inc., Atlanta, GA, 1992.

[4]Fink, J., E. Banaszak, W. Thiede, and J. Barboriak, "Interstitial Pneumonitis Due to Hypersensitivity to an Organism Contaminating a Heating System," *Ann. Int. Med.*, 74, pp. 80-83, 1971.

[5]Rylander, R., P. Haglind, M. Lundholm, I. Mattsby, and K. Stenqvist, "Humidifier Fever and Endotoxin Exposure," *Clin. Allergy*, 8, pp. 511-516, 1978.

[6]Croft, W., B. Jarvis, and C. Yatwara, "Airborne Outbreak of Trichothecene Toxicosis," *Atmos. Environ.*, 20, pp. 549-552, 1986.

[7]Hodgson, M., P. Morey, J. Simon, T. Waters, and J. Fink, "An Outbreak of Recurrent Acute and Chronic Hypersensitivity Pneumonitis in Office Workers," *Am. J. Epidemol*, 125, pp. 631-638, 1987.

[8]Health and Safety Executive, The Control of Legionellosis Including Legionnaires' Disease. *Health and Safety Booklet H.S. (G) 70*, HSE, UK, 1991.

[9]Broadbent, C., L. Marwood, and R. Bentham, "Legionella Ecology in Cooling Towers," *Australian Refrigeration, Air-Conditioning and Heating*, 46(10), pp. 20-34, 1992.

[10]Morey, P., "Suggested Guidance on Prevention of Microbial Contamination for the Next Revision of ASHRAE Standard 62," IAQ '94, pp. 79-88, American Society of Heating, Refrigerating, and Air-Conditioning Engineers, Inc., Atlanta, GA, 1994.

[11]Hugenholtz, P. and J. Fuerst, "Heterotrophic Bacteria in an Air-Handling System," *Appl. Environ. Microbiol.*, 58, pp. 3914-3920, 1992.

[12]Pope, A., R. Patterson, and H. Burge, (Ed.), *Indoor Allergens, Assessing and Controlling Adverse Health Effects*, National Academy Press, 1993.

[13]Morey, P., "Studies on Fungi in Air-Conditioned Buildings in a Humid Climate," *Conference on Molds in Indoor Air*, U.S. Public Health Service (in press).

[14]Morey, P., "Use of Hazard Communication Standard and General Duty Clause During Remediation of Fungal Contamination," *Proceedings of Sixth Intern. Conf., Indoor Air and Climate*, 4, pp. 391-395, 1993.

[15]Block, S., "Humidity Requirements for Mold Growth," *Appl. Microbiol.* 1, pp. 287-293, 1953.

[16]Flannigan, B., "Approaches to Assessment of Microbial Flora in Buildings," IAQ '92, *Environments for People*, pp. 139-145, American Society of Heating, Refrigerating and Air-Conditioning Engineers, Inc., Atlanta, GA, 1992.

[17]Pasanen, A-L, H. Heindnen-Tanski, P. Kalliokoski, and M. Jantunen, "Fungal Microcolonies on Indoor Surfaces - An Explanation for Base-Level Fungal Spore Counts in Indoor Air," *Atmos, Environ.*, 26B, pp. 117-120, 1992.

[18]Morey, P., M. Hodgson, W. Sorenson, G. Kullman, W. Rhodes, and G. Visvesvara, "Environmental Studies in Moldy Office Buildings: Biological Agents, Sources, and Preventive Measures," *Ann. Am. Conf. Gov. Ind. Hyg.*, 10, pp. 21-35, 1984.

[19]Morey, P., and C. Williams, "Is Porous Insulation Inside an HVAC System Compatible with a Healthy Building?," *IAQ '91, Healthy Buildings*, pp. 128-135, American Society of Heating, Refrigerating and Air-Conditioning Engineers, Inc., Atlanta, GA, 1991.

[20]Ahearn, D., D. Price, R. Simmons, and S. Crow, "Colonization Studies of Various HVAC Insulation Materials," *IAQ '92, Environments for People*, pp. 179-184, American Society of Heating, Refrigerating and Air-Conditioning Engineers, Inc., Atlanta, GA, 1992.

[21]Lstiburek, J., "Two Studies of Mold and Mildew in Florida Buildings," *J. Thermal Insul. and Bldg. Envs.*, 16, pp. 68-80, 1992.

[22]White, J., "Solving Moisture and Mold Problems," *Proceedings of Fifth Interm. Conf., Indoor Air Quality and Climate*, 4, pp. 589-594, 1990.

[23]Chartered Institution of Building Services Engineers, "Minimizing the Risk of Legionnaires' Disease," Technical Memorandum 13, CIBSE, 1987.

[24]Morey, P., J. Feeley, and J. Otten, (Ed.), *Biological Contaminants in Indoor Environments*, ASTM, STP 1071, 1990.

[25]Centers for Disease Control and Prevention, *Procedures for the Recovery of Legionella from the Environment*, DHHS, CDCP, 1992.

[26]Morey, P., "Microbiological Contamination in Buildings: Precautions During Remediation Activities," *IAQ '92, Environments for People*, pp. 171-178, American Society of Heating, Refrigerating and Air-Conditioning Engineers, Inc., Atlanta, GA, 1992.

[27]New York City Department of Health, NYC Human Resources Administration and Mount Sinai Occupational Health Clinic, *Guidelines on Assessment and Remediation of Stachybotrys atra in Indoor Environments*, 1994.

Index